# Human Diversity: Its Causes and Social Significance

# Human Diversity: Its Causes and Social Significance

The Proceedings of a Series of Seminars
Sponsored by the
Committee on Human Diversity
of the
American Academy of Arts and Sciences
1973-1974

Edited by
**Bernard D. Davis**
*Harvard Medical School*
and
**Patricia Flaherty**
*American Academy of Arts and Sciences*

**Ballinger Publishing Company • Cambridge, Massachusetts**
*A Subsidiary of J.B. Lippincott Company*

This book is printed on recycled paper.

Published by Ballinger Publishing Company for The American Academy of Arts and Sciences.



International Standard Book Number: 0-88410-047-2

Library of Congress Catalog Card Number: 76-22196

Printed in the United States of America

**Library of Congress Cataloging in Publication Data**

Main entry under title:

Human diversity.

Bibliography: p.
1. Human genetics–Congresses. 2. Human evolution–Congresses. 3. Nature and nurture–Congresses. 4. Race-Congresses. I. Davis, Bernard David, 1916- II. Flaherty, Patricia, 1938- III. American Academy of Arts and Sciences, Boston. Committee on Human Diversity. [DNLM: 1. Individuality–Congresses. BF697 H918]
GN247.H86 573.2'2'2 76-7002
ISBN 0-88410-047-2

# Contents

# Foreword

The relative role of genetics and the environment in determining human capacities and behavior has become a focal point of controversy in recent years. It has become increasingly difficult to reconcile the ethical and political norms of equality with scientific observations of diversity within the human population. Whether this incompatibility is only apparent rather than real is itself a matter for debate, while the data from research in human behaviorial genetics are also subject to conflicting interpretations.

To consider various aspects of human diversity within an interdisciplinary framework, a group of Fellows of the American Academy organized a series of seminars with financial support from the Rockefeller Foundation and from the Academy's Research Funds Committee. It was the hope that private discussions among scholars with a wide variety of viewpoints and from a number of different disciplines could help to sort out some of the underlying scientific principles as well as illuminate the relationship between scientific findings and public policy debates. While no consensus was achieved, the aim of a rational and constructive discussion was realized, and we have been urged to make the proceedings available to interested scholars. We do so in the hope that the publication may suggest to others fruitful areas of research and discussion, and may therefore help lead toward the resolution of issues of fundamental concern to our society.

It should be emphasized that no particular conclusions or recommendations should be attributed to the Academy. In providing an institutional base for inquiries of broad interest to its members, the Academy seeks to encourage the formation of groups of scholars who proceed on their own responsibility to analyze and illuminate relevant aspects of the topic in question. The views expressed are the views of the participants; the Academy itself does not necessarily endorse any conclusions or policy recommendations resulting from such studies.

**Harvey Brooks**
*President*
American Academy of Arts and Sciences

# Introduction

As our society moves toward greater social justice, we are faced with the challenge of dealing with a wide spectrum of differences among individuals and among groups. This diversity is at once an enormous cultural and biological resource and a cause of social and political tension. And though we cannot specify with precision the relative contributions of environmental and hereditary factors to many of these differences, a realistic approach to social policy must recognize the importance of both sets of factors. While legislation and changes in social attitudes can remove many social inequalities—and should be able to eliminate environmental obstacles to individual development—it is not clear to what extent we can hope to modify or to compensate for most hereditary differences.

That the environment has a large role in shaping social and individual diversity is obvious: a person's abilities, drives, and achievements are enormously influenced by cultural traditions, socioeconomic level, education, and life circumstances—including luck. Moreover, social and behavioral scientists have tended in recent decades to rely on environmental factors to explain individual and group differences, partly for methodological reasons and partly because environmental differences appear more susceptible to effective social or political intervention.

The contribution of heredity is less well defined and has received less attention. Behavioral traits are especially difficult to study from this standpoint: they depend on the joint contribution of very many genes; outcomes are the product of the interactions of these genes with the environment and are not determined by genes alone; and we lack direct neurobiological bridges between observations on genes and on behavior. Accordingly, the study of behavior, unlike most other branches of biology and medicine, has benefited little from the recent spectacular success of molecular genetics in probing single genes with an identifiable molecular product.

Nevertheless, modern developments in evolutionary theory and genetics, resulting in the emergence of the field of population biology, have begun to make constructive contributions to social problems. By focusing on the statistical distributions of genes among populations, and on the role of most genes in defining potentials rather than traits, this new field has provided powerful evidence against earlier typological and deterministic misconceptions of race. More specific contributions of genetics are also proving relevant for public policy. For example, the results of an exhaustive study of schizophrenia, summarized in this volume, illustrate the possibility of demonstrating a large genetic component in a behavioral trait and thus providing a basis for approaching the prevention and treatment of this disease more realistically.

But while it is clear that the identification of hereditary differences could have beneficial uses, it is also clear that this knowledge, like all knowledge, could be used in harmful ways. Because of concern over this possibility—or over the current politicization of the topic—many scientists are reluctant to study such problems. Even apart from these inhibitions, biologists and behavioral scientists rarely communicate with each other about such matters, and they may not even understand each other's terms of reference or language. Moreover, the public debates in this area have been mostly superficial and polemical, rather than informed and analytical. We thus see a large area, with important consequences for public policy, education, and medicine, in which there has been no systematic, critical discussion, and in which neither biological nor social determinism is scientifically justifiable or likely to be helpful.

To promote informed communication on these issues the American Academy of Arts and Sciences organized a series of conferences by a group of scholars from various disciplines and with various—and indeed widely diverging—viewpoints. The original plan was to take up the following topics, in order:

1. A review of evolutionary theory, and particularly its implications for diversity within a species.

2. A review of relevant aspects of genetics, including contribu-

tions of molecular genetics to recognition of the breadth of genetic diversity and to understanding of the mechanisms of interaction of genes and environment.

3. The problems involved in measuring and characterizing intelligence and other behavioral traits, and in estimating the contributions of genetic and of environmental factors to individual differences and to group differences in these traits.

4. The implications of current knowledge of human diversity for the psychosocial sciences, including the possibilities for using knowledge of genetic differences in positive ways, and the possible harmful consequences of building policies on false assumptions about the presence or absence of such differences.

5. The question of how public policy can take into account the realities of this biological diversity, given the equally relevant realities of the political process.

We were able to cover the first three topics in considerable depth, but the implications for the social sciences and for public policy were only briefly discussed. It proved impossible to sustain our original intention of avoiding the highly charged subject of specific ethnic differences: after discussing the evolutionary and methodological aspects of diversity the group unanimously requested an opportunity to scrutinize the concrete evidence on variation in IQ, both among socioeconomic and ethnic groups and among individuals. The last meeting was therefore devoted to this topic. In addition, the social scientists did not find it feasible to prepare detailed analyses of the possible implications of the biological evidence for their respective fields. It has therefore seemed advisable to publish the present transcript, in the expectation that the exposition of the biological background, with identification of some of the problems, will be useful for other scholars.

As was expected, the conference group did not reach a consensus on the origins of group differences. They also differed on the reliability of quantitative assessments of the heritability of intelligence among individuals, but they did agree on the existence of substantial genotypic differences. The exchanges were lively and instructive, and it was particularly gratifying to find that in an appropriate atmosphere persons with widely divergent convictions could discuss the topics dispassionately and with increasing respect for each other's views.

The members of the conference group are indebted to Harvey Brooks and John Voss for encouraging the Academy to support exploration of such a controversial and critical subject, and for their personal contributions to maintaining a cordial and open atmosphere. We are also deeply indebted to the Academy's Committees on Research Funds and to the Rockefeller Foundation for financial support and encouragement, and to Patricia Flaherty for converting a taped series of conversations into a coherent, readable volume.

The Chairman is grateful for the hospitality of the Center for Advanced Study in the Behavioral Sciences during the year in which the conferences took place.

**Bernard D. Davis**
Adele Lehman Professor of
Bacterial Physiology
Harvard Medical School

## List of Participants

### MEMBERS OF THE SEMINAR

Daniel Bell — Sociology, Harvard University
Harvey Brooks — Technology and Public Policy, Harvard University
Luigi Cavalli-Sforza — Genetics, Stanford University School of Medicine
Lee J. Cronbach — Education, Stanford University
Bernard D. Davis — Bacterial Physiology, Harvard Medical School
Irven DeVore — Anthropology, Harvard University
Leon Eisenberg — Psychiatry, Harvard Medical School
Yehuda Elkana — History of Science, Hebrew University
Martin Feldstein — Economics, Harvard University
Richard Freeman — Economics, Harvard University
Zvi Griliches — Economics, Harvard University
Geoffrey A. Harrison — Anthropology, Oxford University
Richard J. Herrnstein — Psychology, Harvard University
Philip Heymann — Harvard Law School
Hudson Hoagland — Worcester Foundation for Experimental Biology
Christopher Jencks — Sociology, Harvard University
Seymour Kety — Psychiatry, Harvard Medical School
David Layzer — Astronomy, Harvard University
Gardner Lindzey — Psychology, Center for Advanced Study in the Behavioral Sciences
Ernst Mayr — Zoology, Harvard University
Robert S. Morison — Biology, Cornell University (currently at the Massachusetts Institute of Technology)

Talcott Parsons — Sociology, Harvard University
Lee Rainwater — Sociology, Harvard University
Judith J. Thomson — Philosophy, Massachusetts Institute of Technology
Robert Trivers — Biology, Harvard University
John Voss — American Academy of Arts and Sciences
Edward O. Wilson — Zoology, Harvard University

## INVITED DISCUSSANTS

Mark Adams — History of Science, University of Pennsylvania
George Brewer — Human Genetics, University of Michigan Medical School
Lloyd Humphreys — Psychology, University of Illinois
Arthur R. Jensen — Psychology, University of California at Berkeley
Jerome Kagan — Psychology, Harvard University
John C. Loehlin — Psychology, University of Texas
Steven Matthysse — Psychiatry, Harvard Medical School
Gerald E. McClearn — Behavioral Genetics, University of Colorado

**Committee on Human Diversity**

Bernard D. Davis, *Chairman*
Talcott Parsons, *Vice-Chairman*
Christopher Jencks
Ernst Mayr

Patricia Flaherty, *Staff Associate*

## Seminar I: Evolution
## (October 26-27, 1973)

**MAJOR PRESENTATIONS**

Ernst Mayr: *The Present State of Evolutionary Theory*

Irven DeVore, Robert Trivers, Luigi Cavalli-Sforza: *Human Evolution*

Mark Adams: *History of Controversies over the Social Implications of Evolution and Genetics*

*Chairman:* Ernst Mayr

# Introduction

An opening statement outlining the purposes of the series of seminars on human diversity being sponsored by the American Academy was presented by John Voss, the Executive Officer. He observed that the question of the biological aspects of human diversity has become a particularly sensitive topic in the academic community. For the scientist who is not a specialist in these areas, not to mention the educated layman, this controversy is very confusing. Moreover, it takes place in a wider context of public concern with the issue of equality.

One source of this confusion is that geneticists and evolutionary biologists on the one hand, and social scientists and academics concerned with public issues on the other hand, do not often talk with one another. Thus, a major purpose of these seminars was simply mutual education. Not only do biologists and social scientists need to learn from each other, but geneticists need to be exposed to evolutionary biologists. A second purpose was to stand aside from the public debate over human diversity and to try, in private, and without acrimony, to clarify or at least help define the issues. Therefore, these seminars represent an effort to show that discussion about a particularly sensitive public problem can be conducted in a detached and constructive fashion.

As background for the discussion, Bernard Davis, Chairman of the Steering Committee, described some of the problems relating to the implications of human diversity. He pointed out that the subject has become especially difficult in recent years because our society has reached general agreement that second class citizenship for any group, as well as the existence of poverty in an affluent society, is indefensible. The result has been the development of a wave of concern for social justice and equality that is unprecedented in this country since the crises of the Revolution and the Civil War.

The problem is that the term "equality" is often used loosely, in ways that fail to differentiate social equality from biological equality, i.e., to distinguish areas where society can impose equality of position, rewards, and status from areas where degrees of equality and inequality in various talents and abilities are only partially determined by environmental influences. One result is that genetic variation is widely ignored or even denied in scholarly publications as well as in the popular media. This problem is disturbing to the scientist who would like at least to see the issue defined clearly. More important, it will likely result in public policies and expectations that may well fail if they are partially grounded on myths rather than on reality.

Many thoughtful people recognize this problem but fear that any inquiry at this time would necessarily have a chilling effect on the present war against social injustice. This view is based on a long history of misuse of evolutionary and genetic concepts—most recently by Shockley, who, in trying to advance the study of human behavioral genetics, seriously tainted the field. These meetings represent an effort to show that the traditional dichotomy between liberal environmentalists and conservative hereditarians is too simplistic and is scientifically obsolete: the whole question of nature or nurture is based on primitive misconceptions and must be replaced by the recognition that we are always dealing with interaction between genetic potentials and the cultural influences that do or do not develop those potentials.

Dr. Davis expressed the hope that these seminars will show that biological knowledge need not counter the goal of social justice but can help select effective procedures, based on acceptance of the realities of biological diversity. A strong emphasis in the meetings will be placed on our knowledge of evolution, since evolutionary theory has important implications for the nature as well as for the origin of man; it leads us to expect extensive diversity in man with respect to behavioral traits as well as to physical traits, regardless of the imperfections of methods for measuring heritability of behavioral differences.

## THE PRESENT STATE OF EVOLUTIONARY THEORY

A review of the present state of evolutionary theory, especially in relation to its potential consequences for the psychosocial sciences, was presented by Ernst Mayr. He observed that the first seminar on human diversity is appropriately devoted to evolution because man is a product of evolution, and man is still evolving. The Darwinian explanation of evolution is now the universally accepted theory, but there are many areas where present knowledge is incomplete. These include the relative importance of stochastic phenomena in evolution, including neutral mutations; the possibility of entire breeding populations as units of selection, so-called group selection; the relative contribution of genotype and of environment to important components of the phenotype; the role of population size, isolation, and migration; and the partitioning of selection pressures into the survival component and the reproductive advantage component. However, none of these areas of insufficient understanding invalidates the basic Darwinian theory, which maintains that evolution is the result of two interacting sets of factors, the *ever-present availability of an immense amount of undirected genetic variation*, and the *direction-giving power of natural selection.*

## ALTERNATIVE EVOLUTIONARY THEORIES

It should be emphasized that all other evolutionary theories, some of which were for lengthy periods considerably more popular than Darwinism, have in the meantime been decisively refuted. The first of these relates to the *inheritance of acquired characters*, which was accepted for many years. The decisive proof against it is provided by molecular biology. When the genetic program encoded in the DNA is translated via RNA into protein, this is a one-way street. There is no possible way in which information from the protein can be put back into the RNA and DNA and contribute to the program for the next generation. The protein, which is essentially the phenotype, cannot convey information back into the nucleic acids.

Another refuted concept of evolutionary mechanisms is the idea that there is a *built-in drive toward perfection.* The idea of a teleology of the universe, which was represented in the *scala naturae* of the seventeenth and eighteenth centuries (later called orthogenesis), has been totally refuted. It would be true only if there were a genetic program for evolution as a whole, and this does not exist. The future of biological evolution is to a very large extent unpredictable. One can predict only processes that obey the laws of physics in a deterministic manner. In cosmology certain developments, such as the sequence through which a star goes, are built in, but this is a matter of physical laws that are carried out step by step. It is important to stress the absence of such a regular drive toward perfection in the organic world. In the last century anthropologists believed that cultures went through a regular cycle of ever greater perfection, and when this particular concept was refuted, they thought that this meant that evolutionary theory as a whole was invalidated.

A third refuted concept is the idea that evolution proceeds by large jumps or *saltations*. There is absolutely no evidence for this theory. Moreover, the delicate balance of a genotype, with many genes interacting with each other and with their control mechanisms, indicates the total impossibility of any major jumps.

## SOURCES OF GENETIC VARIATION

Professor Mayr then turned to a consideration of the first component of Darwinian evolutionary theory: genetic variation. For Darwin, genetic variation was a black box: he knew there was abundant variation but he did not know what caused it. New genotypes can arise in two different ways: first, by *mutation* of one of the genes of the genotype; and second, by *recombination*. The new individual inherits half its chromosomes from each parent, and there is a continuous recombination of genetic potentialities in every generation. By far the greatest amount of the observed variability in any species of sexually reproducing organisms is the result not of new mutations but of new combinations of genes. The major role of mutation is that it replenishes the genetic variability of the gene pool of a population, which in every generation is reduced by two processes. The first is stochastic: the loss of certain genes because they happen not to be transferred into fertilized egg cells. The second, far more important, is natural selection, in which certain combinations of genes are more likely to result in phenotypes than other combinations.

## THE NATURE OF NATURAL SELECTION

Natural selection is based on two facts: first, that some individuals will add more genes to the gene pool of the next generation than others; and second, that this differential reproduction is not entirely a random phenomenon—the greater evolutionary success of those individuals that are reproductively more successful is very often caused, or at least influenced, by certain qualities of these individuals.

Most traits exhibit continuous variation. In a majority of cases any deviation from the mean value will be discriminated against, whether it is on the plus or the minus side; this type of selection was known for a long time and is recognized in modern evolutionary biology as *stabilizing selection.*

One of Darwin's early contemporaries used this as evidence that evolution could not take place because any deviation from the mean would immediately be eliminated. Darwin pointed out, however, that there is also *directive selection*: because of certain changes of the environment or other factors, the individuals on one side of the curve could be favored by natural selection, while the ones on the other side would be eliminated and die out. This is the type of selection that leads to evolutionary changes; the other one simply leads to stability.

Seymour Kety asked why a positive trait at the extreme would tend to prejudice against its survival; for example, why would an animal with very great visual acuity be prejudiced against? Professor Mayr pointed out that almost any phenotype that is favored by selection is subject to a balance of opposing selection pressures. For instance, visual hyperacuity might create trouble in the central nervous system. Most genes control several components of the phenotype. They produce a certain enzyme or a certain balance of enzymes that are used in various parts of the organism. If one organ is favored, several of the other organs that depend on the same combination of these genes may be weakened. There are a number of interactions in the genotype and in the phenotype; stability has the highest selective value, unless a character is clearly so superior that the organism accepts the slight inferiorities that may result in other systems.

Dr. Davis added that every cell of a highly differentiated organism such as man has a full set of genes, and that brain, skin, bone, and other cells differ from each other in terms of which genes are turned on and which are turned off. Any one gene very often is turned on not just in one kind of cell but in several kinds of cells, so if a mutation makes its product function better in cell A, there may be some cost in cells B, C, and D. For instance, one could easily design and evolve a man who had a much higher level of resistance to infection than men have today, but it would be at the expense of something else. He noted that when microorganisms are grown over a long period of time in a medium that supplies them with certain foods, they will lose the ability to make that food. It is an important principle that the organism pays a price for any selection; hence, if a character is not useful in a given environment, it is likely to be dropped.

There was considerable discussion of the relation between the curve for the distribution of a trait and its selection curve, which indicates its probability of survival. Luigi Cavalli-Sforza explained that if the selection curve has a peak which is close to the peak of the frequency distribution curve, the result is stabilizing selection. But if the peaks are separate there is directional selection. Dr. Davis emphasized that the incidence of identical curves for fitness and for the actual distribution does not mean that there is no selection. In fact in every generation, as a result of mutation and of recombination, there will be a few more at the extremes; but if the environment remains relatively constant, generation after generation those ends will be selected against and the population will retain about the same distribution of the trait. If the environment shifts, and a greater or lesser amount of the trait becomes more advantageous, selection will favor one end but not the other.

Professor Mayr added that the early Mendelians saw in selection a strictly negative process, which eliminated everything that deviated unless it produced a new species in one jump. However, since by recombination new gene combinations are constantly being tested, and those that are even to a minimal degree superior have a greater probability of being perpetuated, natural selection is in fact a creative process: by constantly testing new combinations it gradually builds up new genotypes and new evolutionary development. It would be quite misleading to express this simply in terms of an accumulation of favorable mutations, although to some extent this is true in microorganisms. In higher organisms it is genetic recombination that is responsible for providing the major source of genetic variation.

Another point to be stressed is that natural selection is a *probabilistic process*, not an all-or-none process except for highly deleterious or lethal mutations. Normally it is simply that some individuals have a higher probability of surviving or of reproducing than do others. That is particularly true for the creative, directional type of selection, which may shift the mean value of the curve of variation.

A further aspect of selection relates to *mortality selection vs. reproductive advantage*. In the classical Darwinian literature the emphasis is on mortality and survival, but that is only one component of natural selection. Since natural selection is measured in the contribution that an individual makes to the gene pool of the next generation, the major factor is not whether the individual reaches reproductive age but rather how many offspring it produces. In the last fifty to eighty years this has become a very important factor in man, since it is obvious that an individual with ten children adds five times as many genes to the gene pool of the next generation as an individual who has only two children. In western society, with the dramatic reduction of child mortality, the selection due to differential reproductive advantage is now greater than the mortality selection.

Professor Mayr emphasized that the *target of natural selection is the individual.* An individual survives, an individual reproduces. The assumption is made that since genes contribute to the fitness or the survival of individuals, this survival of individuals can be used as a measure of the fitness of genes. In theory and on the average this is correct, but the selective value of a given gene may change quite dramatically depending on the particular genotype in which it finds itself. It is quite well known that certain genes which are perfectly viable in the parents are nevertheless highly deleterious in a particular combination. While it is often possible to draw conclusions regarding the fitness of individual genes, the assumptions should not be oversimplified.

Richard Herrnstein and Robert Trivers called attention to the fact that there are cases of group selection where the target of selection is not the individual. It is true that one has to take into consideration the other genes with which an individual gene is cooperating in order to determine how well that gene is going to do, and that fact is best simplified by saying that natural selection works on the individual. But there are cases in which that statement can lead to error.

Professor Mayr observed that one probably cannot fully appreciate natural selection without some philosophical reorientation. At first glance it is quite surprising that two Britishers, Darwin and Wallace, should have developed the brilliant interpretation of evolution by natural selection, when at that time on the continent—in France, Germany, Scandinavia, the Netherlands—there were many more outstanding biologists than there were in Britain. In Professor Mayr's view, the reason was that continental Europe was (and still is) dominated by typological (essentialist) thinking, from which one could not adopt the concept of natural selection as an important process. Essentialism goes back to Plato, who believed that all visible objects of the universe were reflections of underlying ideas (*eide*) which are fixed, permanent, and unchangeable; the observed variability indicated that the visible world consisted of imperfect reflections of this *eidos*. If essentialist philosophy were true, evolution could occur only by saltations that suddenly produce entirely new *eide*; there would then be no natural selection.

Most ideologies have such a typological basis. Racism is an ideology that is based on typology, since it assumes that once a particular *eidos* is named, then all the members of the group that has this name have the same essential character. In the physical sciences Plato's essentialism has some validity: for instance, all individuals in the class of sodium atoms are essentially alike. However, the typological philosophy makes no sense whatsoever with regard to biological populations, where ordinarily no two individuals are identical. The Darwinian theory of evolution is based on an acceptance of population thinking, in which a population is described in statistical terms as composed of uniquely different individuals.

Yehuda Elkana observed that there are many kinds of essentialism and that it is continually changing. It has helped in some fields while impeding progress in others. At the present time a new kind of essentialist thinking—taking processes as the essence of nature—is slowly moving over from evolution in biology to physics. Such a philosophical orientation might lead to looking at elementary particles as elementary processes.

Professor Mayr also felt that the fact that Darwin and Wallace were dilettantes, with no academic training to speak of and with no preconceived notions of what they should be finding, was relevant to their development of a completely new concept. Professor Elkana agreed that the absence of preconceived ideas was useful, but he felt that the question is *why* they chose to look in the particular direction of natural selection. Professor Mayr felt that this was due to the empirical tradition in Britain. He also pointed out that the animal breeders in England emphasized the differences between individuals, while all other cultures (as a result of Platonic essentialism) emphasized the species.

## POPULATION GENETICS

Edward Wilson called attention to a duality of modern evolutionary thinking which is not unlike the duality between wave and particle mechanics in physics. It is true that natural selection operates on the individual. In studying evolution in a single generation at a microevolutionary level, one generally looks at behavioral mechanisms and ecological interactions at an individual level, and one understands adaptation as a response of individual organisms. On the other hand, in dealing with evolutionary time—that is, sufficient time to allow a significant amount of evolutionary change, with easily detectable amounts of change in gene frequency—the individual amounts to nothing, and evolutionary theory is based solely upon conceptions of populations changing through time. It is meaningless to speak of evolution in terms of individual lineages; that notion went out with the demise of Lamarckism. The evolutionist has to think in terms of a gene pool, which is the aggregate of all the genes of a breeding population; evolution has to be described in terms of the changes in the relative frequencies of the genes of interest within that gene pool.

This raises in one's mind the question of the validity and future of conventional or modern evolutionary theory, which is based upon population genetics. The single gene model is the foundation of 95 percent of classical population genetics; yet it is known that genes do not operate simply, and there is great difficulty in conceiving of evolution as the outcome of selection acting on single genes. There has to be a combination of genes, and that is creating great technical difficulty. Professor Wilson suggested that a wholly new picture of organic evolution may be about to emerge, and that this may be one of the turning points in the history of science. It is possible that conventional population genetics based on the gene pool and changes in gene frequencies, and the equations derived from the single gene concept of evolution, might be extended by computers and other means to resolve the present difficulties relating to group selection, the origin of altruism, the observed great genetic diversity, etc. Such a development might make evolution predictive, at least over a short period of time. But in the minds of an increasing number of population geneticists and evolutionists, it is equally possible that one may have to start with something completely new, as yet unconceived.

In Professor Wilson's view the basic Darwinian idea is correct, but the units of selection are at issue. To what extent can one deal with single genes in writing equations of population genetics? Only with equations of population genetics can one predict evolution and make accurate statements about the distribution of characters in populations from first principles. Although these equations are not very adequate—except in laboratory conditions or field conditions over very short periods of time—they are all one has. Even the most intuitive notions of evolution refer back to them through mental or verbal pictures based on them.

Professor Mayr expressed the view that knowledge of the details of evolutionary change in gene frequencies is not essential. In theory, the mechanisms are known; precise calculations of the interaction of genes are not critical. Leon Eisenberg objected that the unqualified acceptance of the evolutionary mechanisms of classical population genetics is unwarranted if new facts are suggesting that the theory may not be true. Professor Wilson felt that efforts at calculation, even though they have become so much more difficult, should be continued. However, while continuing to build or trying to build on classical population genetics, one should be ever ready to abandon it if a genuine breakthrough occurs.

Dr. Cavalli-Sforza emphasized the necessity for further building on classical population genetics; although present knowledge is not sufficient, in some cases it is clearly useful to think in terms of single genes. He described the problems that are encountered in practice. If one looks at a single gene at a time, he ignores two things: first, that other genes contribute to the phenotype, and second, that genes are not segregated one from the other but are present in the same chromosome. It is not true that population genetics ignores these factors, but it is difficult to consider them in a fully satisfactory way that can explain anything. In Dr. Cavalli-Sforza's view, this means that further expansion of the theory is needed, not necessarily that a completely new concept is called for. It is difficult mathematics, but it is being done. Professor Wilson agreed and indicated that his intention was only to raise the possibility that presently accepted genetic theory may be misleading and to urge caution in applying basic population genetics uncritically to human evolution.

## SPECIES AND SUBSPECIES CATEGORIES

Professor Mayr then turned to a description of species and subdivisions of species. The word "species" designates a category that has been defined in many ways. The typologists said that species is that which has a different essence; therefore they used morphological criteria which reflected this essence. This often worked quite well in practice for the taxonomist, but it led continuously to difficulties. It was such superficial criteria that led to the widespread claim in pre-Darwinian days that blacks, mongols, and whites were three different species. The typological definition has been superseded by the *biological species concept*, which defines the species as an interbreeding community or population reproductively isolated from other such communities. The stress is that species consist of populations and that there is reproductive isolation. All human populations are actually or potentially interbreeding with each other—even the most extreme human races, such as Caucasians and Hottentots or Australian aborigines, have interbred without any indication of any reduction of either fertility or viability of the offspring. Hence, if the biological species concept is adopted, there is no doubt that all of man must be considered members of a single species.

In response to Dr. Eisenberg's inquiry as to whether the product of cross-breeding between species is necessarily nonviable, Professor Mayr emphasized that "reproductively isolated" is populationally defined. There may be individual instances of cross-breeding, but there is no intermediate population. When there is a clear-cut intermediate population, that is considered as evidence that the two groups in question are not biological species, because they are connected by that third population with each other. The species concept avoids the sterility-fertility criterion; species can be interfertile under special circumstances (e.g., lions and tigers in zoos), even though in nature they would live side by side without ever interbreeding. Thus in animals the reproductive isolation between species is initially maintained by behavioral mechanisms.

Robert Morison remarked that he did not understand why the rejection of essentialistic thinking in evolutionary biology did not necessitate also the rejection of the biological species concept. Why not adopt a strictly nominalistic species definition if species have no reality?

Professor Mayr replied that the behavior of two natural populations toward each other, that coexist in space and time, provides a completely nonarbitrary species criterion. If such populations do not interbreed we say that they are two species. Their genetic programs provide the isolating mechanisms that maintain them as reproductively isolated populations. Dr. Eisenberg agreed that it is not essentialist thinking to use that definition because it is recognized that under certain circumstances this can be changed, that the definition implies under most conditions or ordinary conditions or conditions of nature. It separates groups that do make sense genetically, because if they don't interbreed, they don't share and redistribute a gene pool.

Dr. Davis pointed out that in nature evolution proceeds through a process of differentiation within a species first into geographically separated subgroups, which are called races, and then into groups that will no longer interbreed when brought together, which are called species. He suggested that it might be useful to distinguish between temporary species, separated for behavioral reasons, and irreversible species, which cannot interbreed for physical reasons. In any case, because the process of speciation creates a continuum, there are bound to be borderline cases, and no classification can evade this problem. The extraordinary thing is that so few are found. If every species is created through a series of steps, why aren't the intermediate steps still in evidence? The answer seems to be that the intermediate stages were not as good, and so they were transitory, while the harmonious combinations became stabilized. Subspecies or races begin to diverge from each other and go through stages until they move so far apart that they cannot mate with each other, first for behavioral reasons and then for physical reasons.

Professor Mayr observed that in a recent paper Jacques Monod distinguished a number of units of evolution. He maintained that the gene is the unit of inheritance, the individual is the unit of selection, and the species is the unit of evolution. Professor Mayr pointed out the justification for describing species as the unit of evolution: as soon as populations acquire the isolating mechanisms they go their own way, fill new niches in nature, utilize different resources, add to the diversity of the locality, and so on.

## DIFFERENCES BETWEEN SUBGROUPS IN THE SAME SPECIES: RACE

Expressing his surprise that the group was concentrating on the taxonomic rank of species, which of all taxonomic ranks is the least arbitrary, Geoffrey Harrison pointed out that a much more debatable question is whether races within the human species clearly exist as taxonomic ranks. There are those who say they do, and that they can be identified and recognized taxonomically; there are those who think they exist but can't be recognized; and there are those who claim that there are no races, only clines (gradients of genetic difference). As background for the discussion of this aspect of human diversity Professor Mayr explained that most species of plants and animals are *polytypic*, that is, consisting of several subspecies or races. Variability correlated with geography is extemely widespread in the animal and plant kingdoms, and the occurrence of races is the normal situation.

There is continuing controversy as to what is the causation and evolutionary meaning of differences between populations of the same species. There are those who say that differences between races are purely accidental, and that no differences of a taxonomic kind below the species rank have any adaptive significance. Darwin and Wallace, however, were convinced that there is a selective significance to differences among different races; and there is a great deal of evidence that much of the observed geographic variation is indeed *adaptive*. For example, species that extend over considerable climatic regions have different physiological tolerances; the differences are not just phenotypic but often have a genetic basis. For instance, populations of animals in deserts are colored like the sand and in arctic forms there is a great deal of whiteness. This does not prove that all geographical variation is the direct result of natural selection. Some of it may be purely *accidental* because some regions were colonized by a group of founders that included only a fraction of the gene pool of the parental population. The new population is then built up on the basis of that limited original endowment, which might by accident be from somewhere near the end of the parental curve of variation. Dr. Davis pointed out that even if it were at the mean, it still might have less of a range of potential variation and hence yield a different population.

Some racial differences may also be due to *sexual selection*. Darwin, who was less aware of the potential climatic selection pressures with respect to differences of human races, thought that sexual selection was one of the major factors in having produced the differences among human races. One has no evidence for this, and in Professor Mayr's view this is a minor factor, though one that should not be excluded entirely. Professor Trivers noted that the reason why Darwin separated sexual selection was that he regarded other kinds of selection as always improving an animal's fit to its local environment. Thus mortality selection is always weeding out those that are less able to survive, whereas sexual selection, which refers to differential access to individuals of the opposite sex, often selected for characteristics that were inversely correlated with survival.

Noting that various recent publications have identified 5, 33, or 56 human races, Professor Mayr maintained that it is probably impossible to define any precise concept of race; about all one can say is that a human race is a human subgroup that has certain characteristics. How many of the differences that are found among human races are due to climatic selection, to the accidents of founder populations, to sexual selection, or to a by-product of selection pressures of another sort (e.g., immunity to certain diseases) is arguable. However, that the majority of these differences were established at one time as the result of selection pressures is becoming more and more accepted.

Judith Thomson expressed the view that it is idle to say into how many categories individuals within the human species should be divided. One can determine what the number of categories should be only in relation to the way in which the classification is to be used. It is important to know to what uses these notions are going to be put, particularly in relation to public policy questions. Professor Thomson suggested that the justification for the notion of species rests not on the fact that a classification is derived from it but that it allows some kinds of predictions about what happens in different parts of the world to different kinds of animals under different kinds of circumstances. There should be the same justification for any classification by race, that it be put to some useful purpose. Professor Mayr noted that classification by race, even though ill-defined, is useful to establish relationships; there are many interesting phenomena with regard to man which the analysis of racial characteristics helps to clarify and disentangle. Martin Feldstein suggested that if biological species are defined by the lack of interbreeding, classification by racial differentiation might relate to different probabilities of interbreeding.

Dr. Davis related the viewpoint of Theodosius Dobzhansky that human races are relics of precultural history. Groups of people who have been geographically separated for many thousands of years, by an ocean, a mountain range, a desert, etc., tend to accumulate extensive differences in their gene pools, including blood groups, certain enzymes, skin color, stature, physiognomy, etc. Where the barriers are easily overcome, as East or West of the Sahara, one can find a gradient or a cline. Today, with cultural evolution, all barriers are breaking down, and people are moving freely between areas where they could not have moved during the tens of thousands of years when the major races were established.

Professor Wilson noted that while it is often thought that subspecies or races are particularly difficult to define in humans, it is just as difficult in plants and animals. The difficulty is symptomatic of the primitive state of speciation theory—if population genetics is still in an early stage of development, it follows a fortiori that speciation theory is even more primitive. The definition of subspecies used throughout zoology is a series of populations that are phenotypically distinguishable, and they are assumed to have genetic differences.

While the disagreement initially was over how much these populations had to be differentiated in order to be distinguished as subspecies (everyone recognizes that that degree is arbitrary), there are three other grave difficulties with defining subspecies in an abstract way. The first is the difficulty in distinguishing populations. The majority of populations cannot be distinguished because the critical parameter, the absence of exchange of genes, is not known. One does not know the rate of gene flow (the flow of individuals who enter and breed in a new population), and one can only guess by looking at the final product, the phenotypic variation. Until gene flow is known it is impossible to draw lines around most populations. The second difficulty, which arises with defining the concept as a geographically circumscribed set of populations, is the clinal nature of variation that occurs in so many characters. It is very difficult to use such characters to draw a line with any precision. The third difficulty is the fact that different characteristics tend to differ in their patterns of geographic variation in a discordant fashion. If one defines a race by one character or by another, one often gets two different sets of subspecies. In Professor Wilson's view, although geographic variation is a strongly marked characteristic of most animal and plant species, the definition of race or subspecies is a typological notion, strongest in zoology or botany because it is a useful means for putting labels on individual specimens or grouping individual specimens, but impracticable as a labeling device for human beings.

Professor Harrison observed that Dobzhansky defined races as populations that differ in gene frequencies. This is quite simple theoretically, but the definition of a population then presents a problem. It is extraordinarily difficult to identify precisely discrete populations; in most situations they don't exist. Also, any two populations of the human species that one defines will differ in gene frequency, so there would be as many races as populations. While agreeing that most of the variation in the world is clinal, Professor Harrison suggested that there are breaks in the cline, quite steep gradients at certain points. Moreover, when one looks at a large number of characters, there is a good deal of concordance as well as considerable discordance. The real justification for racial classification is one of convenience: while there will be inaccuracies, one can define by a single term a large number of populations whose members tend to share many common features.

Professor Herrnstein suggested that a salient part of the question is that groups of people define themselves as races. While there is a general problem of subspecies within taxonomy and biology as a whole, and especially with regard to human populations, another level has to be considered with regard to the human species, because people's view of themselves as members of races has tremendous political and social consequences.

In Dr. Cavalli-Sforza's view, while definition of species is useful because it implies a number of predictions that are usually true, classification by race does not result in anything that is really useful, and in any case races are too fluid an entity to really be defined. He objected to the idea that races are a relic of the past. In part this is true, but it implies that once there were races and now there are none because there has been cultural evolution. The problem is that there is no way to determine when cultural evolution started. There may very well have been a great deal of communication among all members of the human species for a long time, which established the continuity and the clinal variation. Dr. Cavalli-Sforza pointed out that for a million years over all the earth, wherever stone tools were made, it was the same type of stone tool with very minor local variations. It was only later that sharp local races developed.

Dr. Eisenberg suggested that "race" serves for the biologist as a convenient term to explain that what appear to be major physical differences are in fact geographic variations of no great consequence, for whenever two races (e.g., of birds) are brought together they freely produce an intermediate stage, whereas species do not. In contrast, as used in man, the term suggests that the differences are large rather than small: it is invested with a suggestion of other differences that do not follow from the observed differences. Thus in biology race is used to trivialize a phenomenological difference, but in man it has been maximizing a difference that may not mean nearly as much.

Zvi Griliches felt that its degree of significance depends on its informational context. A classification scheme is a method for providing and economizing a certain amount of information. If one is interested in testing for sickle cell anemia, for instance, it is useful to know that the disease is more frequent among a group that from other criteria one can identify as Negro. In his view it would be misleading to trivialize the concept because it does provide information relating to the fact that certain visible aspects of groups can be correlated with other aspects. Professor Mayr emphasized the danger in using the race concept to establish correlations: even if there is a slight correlation it would be so imperfect that it would be meaningless.

Professor Wilson observed that the previous discussion had been dealing with two very different and only loosely related concepts. First, the concept of zoological race or subspecies has a certain utility as a shorthand in talking about the speciation process, with an emphasis on the great difficulty in making precise definitions in the zoological or the biological realm. Second, the utility of the subspecies concept in human beings relates to the predictiveness that can be derived from correlation of characters. Correlations do exist, and as known characters are specified and accumulated the correlations with the remaining ones are tightened. Thus the more the known traits of a given human being are specified, the more precise becomes prediction with reference to unknown traits—which is very useful in a wide realm including medical applications.

However, this genetic correlation of characters is simply a description of one of the results of geographic differentiation in the past; it is not a zoological definition of populational groups. The zoologist wants a shorthand label to put on stages and processes, admitting the imprecision of such labels and the lack of knowledge of the basic machinery. The sociologist, the educator, and the medical practitioner want something predictive to help them deal with individual human beings: information that can be used not to prejudge them but to better their lot.

Gardner Lindzey felt that racial categories, whether broad or narrow, are bad predictors because the degree of genetic overlap between any of them is too great. Professor Mayr pointed out, however, that one knows in what populations to look for certain things like sickle cell anemia or Tay-Sachs disease. Dr. Morison indicated that such correlations as that between sickle cell anemia and the Negro population underline the importance of getting to know the numbers; two or three characters may be highly correlated, very predictive, while others have no predictive value. Dr. Cavalli-Sforza and Dr. Eisenberg observed that the populational distinctions that might in practice be useful from the point of view of situations like screening for sickle cell anemia or Tay-Sachs disease have little to do with the standard racial distinctions. For example, while sickle cell anemia is found in 10 percent of American blacks, in parts of Africa the disease is very infrequent; Tay-Sachs disease is common not throughout the entire Jewish population, but only among Ashkenazi Jews. There is a great deal of geographical stratification of these things, and even the 55 general races would not be at all useful for making this kind of prediction.

Professor Harrison concluded that the race concept is no explainer of anything but rather is something in need of explaining: there is a form of geographical variety distributed over the world; can this be grouped in clusters, and if so, why? The concept does, however, tend to carry in summary form some information about the geographical pattern of variation in the world. For instance, one can identify the geographic origin of a blood sample solely on the basis of its genetic markers, with about 90 percent accuracy. Why such constellations of characters are there is a question for the biologists.

## INTRAPOPULATIONAL DIVERSITY

Emphasizing that every individual is genetically unique, Professor Mayr called attention to the tremendous amount of intrapopulational diversity. Populational diversity can be divided into three subclasses: selectively advantageous, deleterious, and neutral.

The amount of *selectively advantageous variation* in both the human and animal populations is extremely high. There are many puzzling questions in this area: why is there such a large amount of variability and how is it maintained, since natural selection should constantly be eliminating a great deal of it? Part of the explanation is that heterozygotes, which have two alleles (alternative forms) for a given gene locus, are often favored over either of the homozygotes; part of it is due to the fact that when a certain allele reaches a very low frequency it acquires a greater selective advantage than when the same allele exists at a high frequency. In any case, there is no question that a large portion of the genetic variation found in natural populations, including the human species, is apparently of selective advantage.

However, there is also *deleterious variability*. Genes that give rise to genetic diseases are due to a mutation that occurred in an ancestor. These are generally recessive genes, which show up in the phenotype only when the individual has received the same gene from both parents (i.e., is homozygous). Such deleterious genes are accordingly maintained in the population in heterozygotes, who carry the corresponding normal, dominant gene.

A third category, *neutral* genes, have no effect on fitness whatsoever. One of the controversies in the biological literature at the present time relates to how frequent such neutral alleles might be. Some people believe that a large part of the variability found in populations is due to genes that have no effect on viability and therefore are tolerated in the gene pool as different alleles and not eliminated. A considerable number of other biologists think that the number of such neutral alleles is probably very small. However, certain proteins, such as the ones that build up muscle, bone, and connective tissue, seem to be genetically more variable than the ones that deal with important enzymes. So it is possible that in those genes that contribute to structural proteins there could be certain replacements of amino acid residues without any visible effect on viability.

Dr. Davis emphasized the value of diversity within species. In order to survive, a species has to be reasonably well adapted to the range of circumstances, varying in time and place, that it is likely to be exposed to. The presence of a great variety of characteristics insures that some individuals will thrive under one set of circumstances and others under a different set; for example, genetic diversity will enable some individuals to survive a particular infectious disease, while others would survive a different one. In Dr. Davis's view, this diversity in the human species should be built upon rather than minimized. On the other hand, it should be recognized that some genes are purely deleterious. Some people object to any negative eugenics, aimed at eliminating deleterious genes, because that would decrease the diversity of the species. But it is not necessarily desirable to have maximal diversity. Nature has evolved a compromise, with a great deal of diversity within the species but with certain genes being rapidly selected against because they are clearly deleterious. The major aim of current medical genetics, to decrease misery by aiding such selection, certainly does not harm the species—though there may be borderline cases.

## THE BIOLOGY OF HUMAN EVOLUTION

An account of the chronology of hominid evolution and the selection pressures leading to the emergence of man was presented by Irven DeVore. He first discussed the main features of *primate evolution*, indicating that scholars in primatology have an advantage over most students of animal behavior because the living primates provide examples of a wide variety of previous adaptations in the primate order. The primates originated about 70 million years ago among the insectivores. They were a very successful adaptation; in fact there were more genera in the early Tertiary than there are today, when one finds about 50. The primates were very conservative in many ways; for example, they preserved reptilian traits, such as the number of bones in the arm, the position of the bones, the five-fingered hand, etc. The tree shrew is the living representative that links the primates with the insectivores.

A large number of the early primates were olfactorily oriented animals with incisor teeth for gnawing, long snouts, small eyes on the sides of the head, and claws. From the Eocene onwards, and today in all the primate forms, one finds characteristics of a very different sort—stereoscopic color vision, grasping hands, eyes to the front, etc. Early in primate ancestry one finds the development of prolonged maternal care. Among the ground-adapted primates one finds a great deal of sexual dimorphism; there are marked differences between the two sexes, especially relating to robustness. In baboons males have a very important role in defense of the group against predators—one of the more complex adaptations found in the nonhuman primates, in part because baboons are one of the few species that live in multimale, multifemale groups. One particularly prominent aspect of primate behavior, discovered in the past five or six years, is the importance of matrilineal kinship groups, which are found in most group-living primates.

In the course of evolution from tree shrews to modern man certain obvious trends became increasingly important. The first and most important distinctive characteristic of primates was the development of the *grasping hand*, as opposed to the clawed way of moving through the trees. In the prosimians the fingers all moved together; among the Old World monkeys one begins to find a certain amount of precision grip. In baboons it appears to be related to their grooming activity, which is very important to them; it increases in the apes and reaches its apex in humans. The monkeys who arose about fifty million years ago continued the same basic locomotor pattern but with major changes in the senses; that is, the eyes, instead of tending to be toward the side, are toward the front. It is surmised that at that time *stereoscopic color vision* was developed, and the apes and man continue this basic sensory apparatus essentially unchanged.

Professor DeVore observed that these developments seem to be what has been called a preadaptation; that is, the arboreal line provided the hominid, who came many millions of years later, with a precision grip and stereoscopic color vision so that man can perform fine actions in front of a sensitive sensory apparatus with the brain behind it to process the information. The apes introduced another basic locomotor change, that of *brachiation* (arm swinging), which allows great mobility in the arms, shoulders, and wrists. With the hominids another major locomotor change, *bipedalism*, freed the hands for tool use and led to the increase in cranial size.

One notices other distinguishing features in the primates, especially the higher primates—the monkeys, apes, and man—compared to other mammals: they are extremely inquisitive, inventive, manipulative, imitative, and dominance-striving—all of which leads to a special kind of intelligence and makes it difficult for us to even try to measure intelligence in other animals. The higher primates have a diurnal orientation and a sensory perception that is very different from, for example, that of an olfactorily oriented nocturnal prosimian, much less other mammals. Dolphins may well be very intelligent animals, but man would find it difficult to devise intelligence tests for them since they perceive and process the world in a different way.

Professor DeVore pointed out that one finds in the Old World primates all the important social behaviors that are found in humans—deliberate hunting, tool use, etc.—except language-dependent behavior. This gap between human and nonhuman primates, the linguistic chasm, is a major one. For example, chimpanzee communication is very subtle gesturally and capable of expressing many shades of emotion (their greeting behavior is especially elaborate), but vocally it is no more impressive than that of a prairie dog. There is nothing that remotely approaches a phonetic sort of language system, one in which meaningless phonemic bits are combined into meaningful morphemic units.

Professor DeVore then turned to the *evolution of the hominid line*. It is generally believed that the hominids appeared sometime between ten and fifteen million years ago; there is a candidate for a 15- to 18-million-year-old hominid called Ramapithecus, but the fossil evidence is limited to a few teeth and pieces of jaw, and its classification as a hominid is still disputed. The Australopithecine stage of human evolution is very well known, mostly from teeth but also from a large number of skulls and increasingly from pieces of postcranial skeletal material as well. There is a good deal of controversy about these forms, but the predominant opinion is that there are two species (some would say two genera)—a large one, between 80 and 140 pounds, called *robustus*, and a smaller one, on the order of 40-70 pounds, called *africanus*. Representatives of this genus are clearly present in the fossil record five million years ago; it is generally believed that they had split by three million years ago and probably as early as five million years ago.

By about two-and-one-half million years ago, *Australopithecus africanus* had evolved into *Homo habilis*, and eventually reached the *Homo erectus* stage, best known from Java and Peking men, between two million and one million years ago. From about 400,000 years ago to roughly 100,000 years ago there is a gap in the fossil record; the next known stage is *Homo sapiens* and contemporaneously Neanderthal man. Professor DeVore noted that while the Neanderthal has been caricatured by the popular press as a brutish, stooping creature, this in fact has been disproved. His cranial capacity was of the same order as our own, and for that reason many people consider Neanderthal a subspecies or race of *Homo sapiens*. For the period from about 100,000 years ago to 35,000 years ago, there were these two somewhat different hominids contemporaneous with each other, and from 35,000 years ago to the present there is only *Homo sapiens*.

Professor DeVore pointed out that discoveries of the past five years necessitated two important reinterpretations of the story of human evolution. First, until very recently, the Pleistocene was thought to be about a million years old, which would mean that the evolution from the first Australopithecine to modern man took place in a very restricted time period; it is now known, however, that the Australopithecines appeared at least five million years ago. Second, it is now established beyond a reasonable doubt that at least for two million and probably for three million years, two specifically distinct hominids lived side by side in the Pleistocene until the larger one, *robustus*, disappeared, leaving *africanus*.

## SELECTION PRESSURES LEADING TO THE EMERGENCE OF MAN

Professor DeVore emphasized that the development of modern man occurred not in a simple step-by-step pattern but as a result of a *simultaneous feedback system*, of a sort that one presumes is operating on humans today. All the basic characteristics—the manipulation of the hands, increasing cranial capacity, increasing efficiency of bipedal walking and running—affect simultaneously the selection pressures on each other, some of them facilitating those pressures, some opposing them. The result is a complex of compromising selection pressures held in a kind of dynamic tension at all times. For example, bipedalism freed the hands, which allowed the development of tool use; tool use, it is believed, selected for increased cranial neuronal connections and ultimately increased cranial capacity. But then a series of compromises begin.

Bipedalism led to the widening of the female pelvis because the fetus has to turn at 45 degrees or more in coming out of the birth canal, unlike our ape relatives in which it comes straight down the backbone without any awkward curving. Also, hormones that loosen the ligaments in that area at the proper time in pregnancy also tend to loosen the teeth. The infant gets a crushable skull so that it can be squeezed together to get through the canal, and it must be born in an increasingly immature state, which means that it can no longer cling to the mother as do all the nonhuman primates. However, because the mother is bipedal, her hands are free to carry the increasingly immature infant. Yet the trend cannot go too far in this direction; one might think that the female pelvis could continue to get wider to solve some of these problems, but that would begin to interfere rapidly with the mechanics involved in efficient bipedal walking, which started the whole process in the beginning. In any case, the ramifications of bipedalism were exceedingly extensive; one is paying certain costs for it, but the payoffs are clearly very great.

There are two kinds of evidence relating to the evolution of intelligence over the five-million-year period of human evolution: *tools* and *endocranial casts*. First, it should be emphasized that very possibly a great many tools were not stone tools and therefore not preserved. The earliest known tools, exceedingly simple pebble tools—that is, a round pebble with a few flakes struck off—date from three million years ago. These pebble tools persisted for a long time with very few modifications. It was not until mid-Pleistocene times, and in association with *Homo erectus*, that one finds bifacial hand axes. These, too, are quite conservative instruments, and it is not until quite late, the late Paleolithic, that one finds very elaborate toolkits of great sophistication and complexity. Simultaneously one finds the florescence of cave art, and it is a reasonable hypothesis to think of the development of complex tools and of art as being simultaneous with the development of modern languages. Tool use tends to lead to handedness and a division of function between the hemispheres of the brain. This was probably one of the predisposing factors in the development of language, which is centered in one hemisphere. It is believed that Neanderthal, probably *Homo erectus*, and perhaps even *Australopithecus*, had some kind of protolanguage. However, recent studies of the vocal apparatus in Neanderthals indicate that they were not capable of modern speech because of the positioning of the vocal tract and the direction and function of the column of air as it comes up and turns in that region; it is very like the vocal apparatus of the modern human infant but not like that of an adult human.

Cranial capacity is a very crude measure of intelligence. All one can tell from endocranial casts is the outward morphology of the brain. Nevertheless, one sees throughout the living primates and in the fossil record an enormous expansion of the neocortex, and in brain mapping today, compared with living primates, the areas devoted to speech and to tools have clearly expanded enormously. The area of the outside of the brain devoted to the thumb and fingers in humans is a larger area than that devoted to both hands and both arms of the monkey; similarly, the area devoted to the tongue and lips is very great compared to that in the ape brain. Average cranial capacities are as follows: the chimpanzee, about 300 cc; the gorilla, 500 cc; *Australopithecus robustus* and *Australopithecus africanus*, both about 500 (since robustus was a good deal larger, that is an impressive cranial capacity for africanus); *Homo habilis*, 650 cc; early *Homo erectus*, 850 cc, with later skulls about 1,050. By comparison modern male humans average 1,450, and females 1,300.

Popular writers have suggested that since the cranial capacity of Neanderthal and Cro-Magnon tends to be larger than that of modern man, intelligence must be declining. However, one should keep in mind that the cranial capacity of late Paleolithic man is measured on the basis of Cro-Magnon specimens of six feet and more in height; the average for cranial capacity today is a worldwide average, which includes many people of very short stature. The recent literature indicates a high correlation of cranial capacity with stature but not with body weight. Some studies indicating racial or population differences in averages of cranial capacities were brain-to-body-weight rather than brain-to-height measures, and if one corrects for height the differences largely disappear.

It should also be emphasized that not only is cranial capacity a crude measure, but the range between very creative persons in cranial capacity is very great. One uses cranial capacity not as a good measure but as the only measure one has if one is looking at the broad sweep of evolutionary history.

Professor Mayr noted that brain increase at first was quite slow: *africanus* had a brain only about 30 percent larger than that of a chimpanzee compared to body size. Very soon afterward (nobody knows precisely how long ago because those strata aren't exactly dated, but probably about 2.7 million years ago) the brain began to increase rapidly until it reached a plateau about 100,000 years ago. Dr. Davis questioned whether it really has leveled off; 100,000 years is a short period in terms of a scale that covers two million years, and it seems possible, although difficult to measure precisely, that the original curve might have extrapolated through most or all of that period.

Professor Wilson inquired about the validity of the various reasons suggested for the expansion of the brain—whether it expanded as a concomitant of tool using, as a concomitant of the origin of language, or in connection with some unseen development in social behavior. Professor DeVore felt that all three hypotheses are valid. A few years ago the tool hypothesis, which has held sway since Darwin's time, was the most accepted. However, the evidence indicates that the cranial capacity increased faster than the increasing sophistication of the tools, so it is reasonable to view this as the effect of something in addition to tool use.

Hudson Hoagland questioned the validity of the language hypothesis, since Neanderthal had a large brain even though he did not have the facility to use language because of the arrangement in his throat. Professor DeVore pointed out that although there was no language in the modern sense, at least from *Homo erectus* times there had been some kind of protolanguage. Because there is no model for such language today it is difficult to conceptualize, but recent studies in child language learning show the great advantage of even an exceedingly simple set of operation words; the development of a protolanguage similarly would be of great significance.

Professor Griliches asked about the evidence on diversity during the past 100,000 years, whether there is evidence for only one human species, or whether Neanderthal and *sapiens* were separate species and could not interbreed. Professor DeVore replied that although Neanderthal was once viewed as a separate species he is now classed as a subspecies or racial variant by most authorities. Professor Mayr added that when Cro-Magnon entered western Europe about 30,000 years ago the Neanderthal disappeared, but we do not know whether the remnants were absorbed by the invaders or became extinct.

## THE HUNTER-GATHERER LIFE

Professor DeVore then turned to a description of the hunter-gatherer life based on his studies of the San of the Kalahari Desert over a ten-year period. Men have been hunter-gatherers for two or three, possibly as long as eight million years. The invention of agriculture and pastoralism took place only 10,000 years ago, and as recently as 2,000 years ago probably half or more of the world still lived the hunter-gatherer way of life.

If one describes more than the three basic races—black, white, and yellow—the next level of criteria would probably create an additional two, the Bushmen of the Kalahari and the Australian aborigines. The Bushmen, or the San as they are now called, have characteristics that are, presumably through parallel evolution, mongoloid-like, but they are basically a subcategory of African peoples. There is a good deal of indirect evidence that they are the original inhabitants of Africa; the art associated with them in historic times is found in the Sahara and throughout all of savanna Africa. The San live in tightly clustered camps, with very low population density—of the order of one person per three square miles. They are short—the men about five feet, the women two or three inches shorter.

Professor DeVore's studies of the San indicate that the role of gathering has been greatly underestimated. There are long periods of time when the women, who go out in the early morning and gather for several hours, contribute 80 percent of the food by weight and by calories to the diet. Hunting is a very risky proposition for human hunter-gatherer groups, and they would lead a very hard life if they did not also have the food provided by their gathering activity. The San work only about twenty hours a week and therefore have a great amount of leisure time in which they dance, create beadwork, play musical instruments, and play games.

Professor DeVore enumerated several distinguishing characteristics of the hunter-gatherer way of life. The first is a high degree of *sharing and economic reciprocity*—a characteristic not found in any nonhuman primates. This is not to say property is communal; there is very definite ownership, but there is also sharing and the expectation of sharing, and in Professor DeVore's view this is the heart of the hunter-gatherer adaptation. Also, the *population is very fluid*; they do not move in bands that are tightly structured and immutable, but rather individual families are constantly moving, literally almost daily, from one cluster to another, and attaching themselves to these basically by kinship ties both by blood and by marriage. The family moves independently of other families, but that does not mean it lives outside a social network. Their independent nuclear family organization may be due to the fact that they are living in a marginal environment and must constantly move around to follow the resources. This assumption might help to explain the curious fact that a bilaterally extended kinship group with independent nuclear families is found at the two ends of the socioeconomic spectrum, among hunter-gatherers and among ourselves, since in modern urban society one is also following the resources. The vast majority of human societies, in contrast, are based on matrilineal or patrilineal blood lines. The nonhuman primates also have one form of organization based on pair bonding, and another based on matrilineal kin lines.

Conflict is resolved among the San by fission; with their extremely fluid movement and friends scattered in many places in different nuclei, when trouble arises one of the two families involved will simply move. There is no overarching political authority, not even a head man, so conflict resolution by fission in the group is very important. If a chronic problem develops there is murder and blood revenge. In fact, although the frequency is not high, because the population is small, the murder rate is the same as in the United States, probably for some of the same reasons: they carry dangerous weapons with them and they have a male honor orientation.

The available genealogies, collected on about 3,000 individuals, indicate that the San population is in equilibrium, provided one makes certain assumptions. The birth spacing is on the order of three-and-a-half to four years. Most of the time this occurs naturally. There is no explanation, but because of the lack of supplemental foods, nursing continues over a long period: the last child may be nursed until it is four or five years old, and this appears to suppress ovulation. Infanticide is practiced if the next child comes before the preceding one could be weaned, which is a spacing of about three or four years. The mother simply cannot cope with two children, since she must carry the child with her during the day. This carrying of the infant seems to be the major limitation—as soon as the San settle in the acculturative villages birth spacing drops to two years, and the population chart goes up far out of proportion to what one would expect on the basis of the food value of the relatively small increment of milk they receive.

The phenomenon of African precocity has for many years been the subject of intensive investigation. One of the earliest studies was done in a hospital in Kampala, where newborn infants of Asian,

European, and Bantu mothers were given various tests and the African infants showed a significantly more precocious performance. Professor DeVore's group carried out a number of tests on young San infants, which showed no significant differences from European controls. On the other hand, tests a few months later showed quite significant differences, at least some of which appear to be directly related to child rearing attitudes and practices. The San believe that the child should be helped over certain developmental milestones—walking, sitting, and so on—and the tests showed that they can perform these functions better than European infants of the same age. However, what is more interesting is their ability in the succeeding months to do better on a series of cognitive tests. The primary reason may be that the San infant is always held in a vertical position. Tests on European and American controls were always administered in the horizontal position, but it was found that if American infants are raised to the vertical, their performance approaches that of the San.

In response to Professor Griliches' question as to what kind of variance one finds in the population, Professor DeVore indicated that anthropometrically the San are more homogeneous than a population drawn from Boston or New York, which is not surprising since so many populations have combined in the American cities. However, one finds the same variety of personal characteristics that can be found in any urban population: some are dull, some are bright, some are very creative, and so forth.

Professor Thomson asked Professor DeVore whether the results of his studies would tend to support one moral, social, or political stance or policy over another. Professor DeVore pointed out that his work was undertaken not for the purpose of developing social policy but as an objective scientific inquiry. On the basis of what has been analyzed so far, if a population of San were settled in an urban environment at the present time one might expect them to perform in school about as well as a cross-section of any other population. There are, of course, populational differences between them and ourselves, but the range that they show overlaps so significantly with our own that there would be no basis for assignment except as individuals.

Dr. Morison felt that one major difference is that they have managed to keep their population density down to one person per three square miles, and that they do this partly by nursing their children for four years and partly by infanticide. In his view, if a moral lesson is to be drawn, it would be that abortion is a good thing. Professor Trivers maintained that one cannot infer that those practices are what controls the population. There is no evidence that they are not trying to reproduce as fast as they are capable; in fact, birth spacing increases the effective reproductive rate, because healthy infants have a better chance of surviving than infants born close together and less well fed.

## SELECTION PRESSURES IN MAN

From the standpoint of classical evolutionary theory natural selection favors traits that allow the *individual* to leave more surviving offspring. However, Professor Trivers pointed out that the evolution of social traits involves individual behavior that benefits the *group*, without necessarily maximizing the individual's direct offspring. Among langurs (a primate species) in India adult males often kill infants. The initial interpretation of this activity was based on species advantage: it presumably benefited the species by preventing extinction by overpopulation and depletion of resources. However, the behavior has two features that rule out this interpretation. First, it has been observed at low population densities. Second, the behavior would not provide an efficient population regulation mechanism: it occurs only during the first six months after a male ousts the former dominant of a male group, and only with infants less than four months of age.

This behavior is in agreement with the classic process of natural selection: the male finds a series of females who are nursing young infants who are not fathered by himself; in an effort to maximize the representation of his own genes in the next offspring, he kills such infants, which brings their mothers into immediate estrus. He does not kill infants older than six months because the mother is about to cycle anyway; and he stops this kind of behavior when his own children are beginning to be born. This is a striking example of a selfish behavior pattern that confers a strong increase in reproductive success on the male at a cost to the female and infant, and they are unable to do anything about it.

Dr. Eisenberg asked about the nature of the behavioral control mechanism that enables a male langur to destroy the young for the first several months when under other conditions male langurs are fairly tolerant of intrusions on themselves by the young. Professor Trivers replied that nothing is yet known about the mechanism. Moreover, the male langur, in contrast to other primates, is in fact not very tolerant of infants at other times. Professor DeVore added that infanticide in primates is very specific. That is, in a large number of species if a male encounters a strange female who has an infant he will almost invariably start attacking the infant. However, if he has been living in the group for some time and is the dominant male he is very paternal, which is consistent because he is likely to be the father. Dr. Morison asked whether there is any evidence for a pheromone that identifies the infant to the male as his offspring. Professor Trivers replied that while there is no presently available evidence, it is not impossible.

Professor Griliches questioned the statement that the langur's behavior represents an attempt to maximize his contribution to the gene pool. In his opinion, that conclusion is as teleologically unsatisfactory as the view that sees it as the maximization of the species' survival. He suggested that the behavior could be better interpreted as a mechanism to maximize frequency of coitus, which leads to the maximization of his contribution to the gene pool. Professor Trivers stated that in primates the dominant males compete with each other not to maximize their copulations but to maximize their copulations at the time in the estrus cycle when the female is ovulating, which corresponds with paternity. Professor DeVore added that the dominant males in baboons and macaques seem oblivious to a fully estrus female until something (probably an olfactory signal) triggers a reaction. Less dominant males are allowed to copulate at other times.

Professor Trivers proceeded to point out that Darwin explicitly rejected species-advantage interpretations, believing that if one could identify characteristics that harm individual reproductive success but help the species it would invalidate his theory. However, when evolutionists became concerned with social traits, they encountered the problem of altruism. This is most dramatically illustrated by the social insects, whereby one harms his own reproductive success to aid that of another individual. This apparent contradiction to the theory of natural selection was a major factor in the acceptance of species-advantage thinking, which has been a dominant mode in the social sciences and even in biology.

In the past ten years it has been increasingly agreed that species-advantage thinking is erroneous. One alternative model to explain selection for cooperation is the theory of degrees of relationship or *kinship*: this is the only emendation of Darwin's notion of natural selection that is generally accepted. The argument, as first stated casually by J.B.S. Haldane and later in some detail by W.D. Hamilton, is very simple: if there is a gene in one person for an altruistic act towards another individual, and if that individual is related to the first one, then there is a specific probability that both have the same gene. If the benefit that one gives to the other is large enough compared to the cost he inflicts on himself and the probability that the other has the same gene, the gene itself can show a net benefit.

In order to make predictions from this model one has to calculate degrees of relationship, which is very simple in diploid species such as man. They follow automatically from two principles of genetics: that half the genes come from the mother and half from the father, and that a parent has a 50-50 chance of passing either member of a pair of homologous genes to any child. For example, for full siblings the degree of relationship is one-half; for half-siblings, with one common parent, it is one-quarter. Along with indicating the conditions under which altruistic behavior will evolve, kinship theory also explains the conditions under which selfish behavior will be opposed, since there are some kinds of selfish acts that one would not want to perform because they would inflict too great a cost on another relative. For example, if the male langur coming into the group would be killing his cousin's children, it may be that the cost is too great compared to the benefit and he elects not to do it.

There are at least three reasons why kinship theory is of critical importance. First, it appears to apply to many known classes of altruistic behavior. For example, in the social insects the sterile ants and bees are helping their mother to produce their sisters and brothers. Moreover, there are other details of ant, bee, wasp, and termite family life that are predicted by this theory, and by none other, from the unique degrees of relationship that occur in these species as a result of their special genetic system, called haplo-diploiding.

Second, kinship theory provides a formulation that permits evolutionary biologists to study relations within human families in order to determine whether one can predict when to expect competitive interactions between siblings, when to look for cooperative ones, when parent-offspring conflict will arise, etc., by working out the argument in terms of degrees of relationship. This was impossible until one had formulated a way of dealing with interactions between closely related individuals.

Third, kinship theory solves an ancient problem in social theory—the extent to which the individual organism is a legitimate analogy to a social group. A biologist looks at an individual organism and sees a hundred billion cells, each of which, since they are identically related to each other, is willing to forego a benefit as long as a neighboring cell receives a greater advantage; one expects no competition within the human body over allocation of resources. It is clear that any social group that does not consist of identically related individuals is necessarily going to be disanalogous to a single organism, since one expects a certain amount of disagreement and conflict from the fact that the individuals are not identical and therefore may not always see situations in the same way. One can now indicate the extent to which the analogy is legitimate as a function of degree of relationship. Insect colonies can almost be treated as if they were a superorganism because degrees of relationship are exceedingly high; however, they are not perfectly identical, and one finds some conflict.

A second explanation for altruistic behavior involves the *possibility of return benefits*: if A performs an altruistic act toward B, who in turn performs it toward A at another time, the resulting advantage could be subject to natural selection. The problem is that the favor does not necessarily guarantee reciprocation. One can develop a detailed form of this argument in which human emotions would be set up in such a way that they end up conferring mutual benefit. One can then derive the hypothesis that natural selection could have favored a sense of justice, a sense of concern over fair apportionment of costs and benefits in a system of reciprocal exchanges. Thus even characteristics that have seemed clearly to result entirely from cultural development could have genetic components as well.

While there has been some difficulty in reconciling altruism with classical evolutionary theory, there is an even greater problem in reconciling intraspecies conflict with species-advantage thinking. If one interprets behavior in terms of helping the species to survive, why is there disagreement and conflict within a species or within a family? Some social scientists, implicitly thinking in terms of the species, explain conflict within species as a mechanism for social change. An alternative, anthropological view is that conflict between parent and offspring during socialization helps the offspring to develop. Professor Trivers suggested that in both cases conflict is more immediately and better explained by simple reference to competing genetic self-interest.

Kinship analysis is particularly important to humans for two reasons. The first is the lengthy period of parental care characteristic of humans. The second relates to the evolution of male parental investment, which is a critical change in hominid evolution. In all human societies males contribute to the care and rearing of their young. This tends to increase the degrees of relationship in local groups, because along with male parental investment comes a strong tendency towards monogamy or polygyny, with all of the female's offspring being fathered by the same male.

Professor Wilson pointed out that the male-female bond in human societies is unique in occurring in large societies. Other societies with a close male-female bond (the gibbon, the marmoset, etc.) are really just family groups. Professor Griliches noted that the theory of kinship implies some mechanism for the recognition of degrees of kinship and would fail where that mechanism does not operate; he inquired whether anything is known about the nature of that mechanism. Professor Trivers replied that while little is known about such mechanisms, in the social insects there are pheromonal indications for membership in the same nuclear family.

Professor Trivers called attention to an irony that follows from the selection for altruism. As the commonality of interest within a group is increased, either because of closer genetic relationships or because of a system of reciprocal benefits, intergroup conflict becomes more likely. In lower species, with less sharing, only the most dominant individuals will benefit from the conflict, and so the subordinate individuals will not be highly motivated to suffer the costs involved in fighting another group. However, as far back as one can trace historically, intergroup conflict has been a regular, almost predictable feature of human social relations. This is consistent with natural selection. It is intriguing to speculate that the development of tools, and the subsequent advances of technology, introduced an automatic instability in intergroup relationships in humans which accentuated selection pressures for populations prone to intergroup conflict; the relationship between technology and warfare in the world today is obvious. Dr. Hoagland asked whether man is the only primate that engages in lethal group combats. Professor Trivers replied that while other primates do have such conflicts, it is not possible to determine whether their murder or death-from-conflict rate is lower than ours; it is clear, however, that there are species which are more murderous.

As mentioned previously, Darwin supposed a greater relationship between sexual selection—that is, competition within one sex for access to members of the opposite sex—and the differentiation of races than is presently accepted. Nevertheless the argument for sexual selection itself is important, and Professor Trivers reported that there has been considerable progress in clarifying its relation to the evolution of sex differences. One critical variable is relative parental investment—how much the father invests in a given child compared to the investment of the mother, investment being defined as anything that helps the child's chance of surviving. For most mammals relative parental investment is zero; the male invests only the sperm while the female invests enormous resources. This variable controls a series of other variables, because members of the sex that invests less will be able to produce more offspring. Since the relative parental investment must be controlled by natural selection, an interesting area of research will be to determine why some birds, fish, and insects—as well as humans—have increased male parental investment from its usual level of zero, the sperm cell, to something quite substantial.

The explanation for sexuality itself must relate to the diversification of offspring in order to deal with unpredictable future contingencies. Some species alternate asexual and sexual phases, and a number of these groups appear to reproduce asexually when conditions are relatively static and sexually when they have to disperse their offspring widely to unknown environments or when the environment appears to be changing. Sexual and asexual reproduction represent two mechanisms for regulating diversity of offspring versus close relationship to the offspring. Inbreeding and outbreeding regulate it in the same way: if, in species in which the males invest nothing, a female decides to let her brother fertilize her offspring, she increases her own degree of relationship to her children from one-half to three-quarters.

Professor Griliches inquired whether male dominance is prevalent in the primates. Professor DeVore replied that it is nearly universal. The nearest approach to equality in the sexes is in strongly pair-bonded groups such as marmosets and gibbons, where there is a good deal of male parental investment. It is interesting to observe that humans have the same degree of sexual dimorphism in terms of robustness, weight, stature, etc., as chimpanzees, who are very male dominant.

Dr. Davis asked whether anything is known about the nature of the mating patterns that contributed to the rapid selection for an increasingly complex brain over the last couple of million years. Professor Mayr replied that there has been some speculation that earlier mating patterns had a bias in favor of the most dominant males, which might have speeded up the process of selection significantly. As indicated earlier, there was a two-million-year period during which the brain doubled or perhaps tripled in size, after a ten-to-twelve-million-year period during which there was little change. The question is whether this could have occurred if every male had had an equal chance to reproduce. In a number of recent primitive human societies certain esteemed males had several females while other males had none. If such a pattern had been prevalent during the period of the rapid increase in brain size in man it would help to explain that evolution.

Professor Wilson called attention to another piece of speculation:

the hypothesis that group selection, through warfare between groups (genocide), may have been critical in speeding up evolution. The possibility is increased in human evolution: once the species crossed a certain threshold of intelligence its members became conscious of, and could discuss, their status and that of groups around them, and it then became increasingly possible to mount organized campaigns of intrusion, dominance, and even genocide. Once that started it could have been an autocatalytic process, and it would have very strongly favored the increase not only of intelligence but also of altruism within the group.

The obvious reference one can make to human society is the altruistic nature of young male warriors with reference to their own lives. If one checks the outlines of small nations and tribes in the Mediterranean area since the earliest historical times the result is astonishing: tribal groups continually become extinct or are absorbed. This could reflect group extinctions that are more than high enough to drive up the frequency of altruistic genes.

Professor Mayr indicated that in most primate societies, even though the group has a certain amount of cohesion, individuals occasionally move to other groups, and this pattern was probably also prevalent in the hominid ancestry at one time. But with more and more organized hunting intragroup communication patterns arose that would not work well with other groups; this would intensify group selection. Moreover, a group that had a favorable new invention, such as a new agricultural product, weapon, or strategy, would benefit. This is a unique situation not found in other parts of the animal kingdom.

Professor DeVore questioned the altruism associated with young male warriors, pointing out that the path to status for males was to be valorous in battle, and their payoffs after the battle were large. Once this quality is deemphasized as a status characteristic, males have to be conscripted to go out and fight. The smaller scale societies are really extended kin groups, and where altruistic sacrifices are made at times, they can be explained by the close kinship ties. Early human societies are based on intensive kinship relationships. Martin Feldstein asked how close the kinship relation must be to be of behavioral significance. Professor DeVore observed that when the kinship is close one can specify the situations in which one would expect conflict and those in which one would expect cooperation. For example, in many African societies there is intense conflict among half-siblings within the local kin group, but they band together immediately against the next group. As the relationship becomes more distant this behavior becomes increasingly difficult to measure.

## CHANGING PATTERNS IN MAN

Citing statistical data on mortality and reproduction, Dr. Cavalli-Sforza called attention to the fact that in the western world there has been a very great *increase in life expectancy* during the past 250 years. The probability of survival from birth to age 15 had been relatively constant at about 50 percent until it began to rise about two or three centuries ago; now the probability that modern western man will reach age 15 is nearly 100 percent, and there has been a similar increase in the percentage of survival to age 50. Hence most of the forces of natural selection through differential mortality have disappeared in modern society. With this increase in survival, there was a *decrease in fertility* but only after a lag of one or two generations. This *demographic transition* is almost complete in western countries, but in the developing countries the process is still in an early stage.

If prereproductive mortality is abolished, what natural selection remains, at least in western populations, is due to differences between families in the number of children. If birth control expands and should lead to not only a smaller number but also a constant number of children, evolution by selection would be abolished.

Professor Harrison asked about the effect of different generation times (i.e., mean age of parents at time of birth of offspring). Professor Wilson observed that since selection is based on relative net growth rates, which are computed by fertility schedules as well as by mortality schedules, increased fertility schedules will result in higher net growth rates. Dr. Cavalli-Sforza noted that if shorter generation time is a biologically determined trait there will be a selection in its favor, which would proceed according to the standard rules of natural selection. As a consequence, other genes in the same chromosome close to the gene that determines lower generation time would be selected at the same time, because in crossing over they are not easily separated. Some diseases may be more frequent in some populations because they have been "hitchhiked" in this way by a gene that has an advantage.

## ASSORTATIVE MATING

Dr. Cavalli-Sforza observed that assortative mating is sometimes called positive or negative, depending on whether it favors like or unlike mates. In general it is positive. For many genetic traits, such as blood groups, mating is strictly random, but for others there is a distinct assortative mating pattern. The correlation coefficients obtained in a number of studies for various physical traits are weak—

mostly between .1 and .2—but they do exist. Dr. Cavalli-Sforza pointed out, however, that these data are simplified; all the studies were done without eliminating social stratification. It is known, for instance, that there are important social stratifications for stature, but nobody has yet done partial correlations to determine how much of the assortative mating for stature is direct and how much is secondary to assortative mating for social class.

Professor Griliches felt that there was a problem of interpreting the mechanism that produces the observed correlations: they may be due not to direct assortative mating but to geographical segregation of groups. Dr. Cavalli-Sforza stated that much depends upon the sample; in general, most marriages tend to take place between people who are born relatively close geographically, and the phenomenon is accentuated in rural areas.

Dr. Eisenberg pointed out another problem relating to the data on correlations in a physical trait like stature: if all people had optimal nutrition (so that any difference in height was genetically determined), so long as tall marries tall the children are tall. However, since these studies did not take social stratification into account, the sample is likely to include people whose genetic possibility to have been tall was concealed by lack of proper nutrition, and they will be passing on genes for tallness even though they measure short. The phenotype is then not representative of the full possibilities of the genotype. Therefore, the correlation will be higher for a group of middle class parents than that for a group which cuts across the whole social spectrum.

Professor Mayr observed that homogamy (positive assortative mating) was taken very seriously in the last century because of the belief that it could lead to the breaking up of the human species into several species. Since it is now known that homogamy does not change gene frequencies per se there is much less interest in the subject. However, Dr. Davis noted that homogamy does broaden the distribution curves, which has considerable social importance.

Dr. Cavalli-Sforza indicated that the data show a relatively high correlation between husbands and wives for IQ, an average of .5. For any modeling of what might happen in the future it is important to identify the source of this correlation. The models usually employed are unrealistic, in the sense that they assume that the correlation is between the genotypic values of the IQs. However, only the phenotypic values are known, and it is not easy to make genetic predictions from correlations on these values. A high correlation for educational attainment has been observed (.7), and in Dr. Cavalli-Sforza's view the source of the homogamy for IQ is essentially social—that is, people usually meet each other either in school or in jobs.

## RELAXATION OF SELECTION PRESSURES

As a result of the decrease in natural selection among civilized populations, certain genetic defects may be becoming more prevalent. Dr. Cavalli-Sforza described several characteristics which appear to have resulted from the relaxation of selection pressures: colorblindness, impaired visual acuity, and incapacity to lactate. It is known that the incidence of colorblindness in populations of hunter-gatherers is lower than in civilized populations. Some have suggested that colorblindness for a hunter-gatherer is such a severe handicap that it will be selected against, while for a farmer or a person in urban society it is not so important. However, there is a serious problem: the only thing that can change the frequency of colorblindness, once selection ceases to operate, is new mutations, and the mutation rate necessary to raise the frequency from one level to another in a given time can be calculated. Since this calculation results in a mutation rate that is much too high, it may be that the cause is not (or not only) relaxation of natural selection, but something more complicated.

Measurements of eye refractions show that in primitive populations which lack eyeglasses deviation from optimum is relatively rare; there is a considerably greater amount of both myopia and farsightedness in civilized populations. A trait like this is different from colorblindness in that for colorblindness only one or two genes are operating, while for a character involving the shape of the eye there are probably a great number of genes; thus the total mutation rate may be much higher. The assumption is that with the transition to agriculture the selection for good vision was relaxed because it was less important than in the hunting and gathering life.

The tendency to lose the capacity to suckle children may also be a case of selection relaxation after the development of agriculture made milk available from other sources. A study of British women indicates that 5 or 6 percent could not lactate properly, as compared to zero among primitive populations. Dr. Eisenberg called attention to the necessity of determining whether the diagnosis of hypolactation in a civilized population really reflected a genetic defect or might be socially induced, e.g., by tension or anxiety. Dr. Cavalli-Sforza indicated that the statistic was claimed to represent genetic hypolactaters, although substantiation for such a claim might not be entirely satisfactory.

It is clear that the introduction of agriculture had tremendous effects on human evolution. This cultural development presents the same question that one finds in genetics: whether a mutation arose in different places or occurred only once and spread to different areas.

Dr. Cavalli-Sforza indicated that archeological knowledge suggests that there were several independent centers of origin of agriculture in the world. The centers that are known—the Middle East, central Africa, Mexico, the Far East—were so remote from each other that even if they differed by several thousand years in the time of the emergence of agriculture, it appears unlikely that there had been communication. Also, these centers produced entirely different crops, and the civilizations that arose varied greatly. The archeological data show that the spread of each civilization from the center has been fairly regular in all directions. The European data, which are numerous and accurate, indicate that the rate of spread has been about one kilometer a year, with local difficulties whenever there were geographical barriers.

A problem of interest from a genetic point of view is how agriculture spread from these centers: it might be a matter of ordinary cultural diffusion, in the same way that a new technique is seen and adopted and carried on to another place; or the people themselves might have spread. Although the first model is generally accepted for most cultural diffusion, it may not be the mechanism of agricultural diffusion. For agriculture allowed population densities a thousand times higher than those of hunter-gatherers, and with such a difference in carrying capacity there must have been periods of rapid growth, during which people could expand from the centers of origin of agriculture. This would mean that Europeans, Indians, and northern Africans originated in the Middle East. Later movements of people blur the picture, but this hypothesis is in agreement with the fairly similar genetic composition of these populations.

Professor Mayr cautioned against ruling out entirely the possibility of agricultural expansion by the usual process of cultural diffusion. In New Guinea he found that the local tribes were inveterate exchangers of things, and some of the spread of agricultural inventions may well have been due to this kind of trading. Dr. Eisenberg added that some cultures readily accept new ideas and methods while others do not.

## IMPLICATIONS OF EVOLUTIONARY THEORY FOR HUMAN GENETIC DIVERSITY

Dr. Davis pointed out that a certain amount of skepticism about the validity of evolution has continued among some groups of intellectuals, as well as much of the general public. With the development of molecular genetics, however, an exceptionally direct kind of evidence has appeared, in which the degree of similarity of DNA sequences provides a rigorous quantitative measure of the degree of evolutionary relatedness. (This evidence will be reviewed in the next seminar.) Since evolution has important implications for the origin, nature, and distribution of human diversity, and since this diversity is central to some serious social problems, it seems important to encourage more extensive education about evolution.

It is remarkable that on the basis of formal relations, with so little knowledge of the mechanisms involved, Darwin was able to infer so much that was supported by subsequent discoveries. Similarly, atoms were inferred on a formal basis long before radioactive decay provided direct evidence for their existence; and genes were inferred long before their structure and their functions were revealed by molecular genetics. Dr. Davis pointed out the similarity of these situations to the problems now being faced in examining the nature of diversity in man. The lack of precise knowledge about the nature of human variation does not mean that one has no knowledge. Nevertheless, regularly in the news media, and often in the scholarly literature of the social sciences, one finds statements that completely ignore the possibility of genetic factors in areas where the interaction of these factors and the environment is important. For example, innumerable articles imply that if the childhood environment is optimized all problems of defects in educational performance will disappear.

While estimates of the heritability of IQ vary, in Dr. Davis's view one cannot ignore the existence of a significant genetic component without serious consequences. Professor Harrison cautioned that one should keep in mind that the problem is not only disagreement on heritability estimates; they are not absolute in any case. Dr. Davis agreed but stressed the point that if one disregards the genetic component in behavioral traits because it is not absolute, or because it cannot be measured precisely, one is in effect working with an assumed value of zero, for which there is no scientific support. Recognizing genetic diversity should not discourage us from trying to optimize childhood environments; rather, it should make our efforts more effective.

Dr. Davis proceeded to point out some conclusions that might be drawn from the previous presentations. The question of the existence of a wide range of genetic variance in behavioral traits can be separated from the question of how precisely we can measure either that variance or the genetic capacities of an individual. Evolutionary theory would bear on the question of the existence of such variance even if no one had measured an IQ. In contrast to the earlier typological approach to species and races, which implied that every such group has one normal type around which there is a certain amount of accidental variation, Darwinian theory leads to the conclusion that every subspecies that has long been isolated and subjected to different selection pressures is bound to accumulate genetic differences. In the hominids the strongest selection pressures over the last two million years have been for increasing brain size and complexity of the nervous system, with a resulting exceptionally rapid and extensive differentiation of this species from its precursors, and from other species, with respect to intellectual capacities. It seems reasonable to conclude that such powerful selective forces would also be at work in subgroups within our species that have been separated for long periods of time.

In other words, from what is known about evolutionary theory it would be realistic to expect statistical genetic differences among different races in behavioral as well as in physical traits, some being selected for advantageous traits in one direction and others in another. Though even larger differences in behavior have been created by different cultural traditions, differences in the distribution of capacities become increasingly important with increasing contact and migration of people from different parts of the world. One of the major themes in our later seminars will be the implications for the social sciences and for public policy of the statistical differences (with extensive overlaps) that we can expect to exist between human subgroups. We must try to utilize this diversity in positive ways, and with compassion—but we cannot solve the problem by willing its disappearance.

Professor Mayr expressed the view that in dealing with the evolution and meaning of human diversity in these seminars it would be more meaningful to concentrate on individuals rather than on subgroups in the society. Dr. Davis agreed but pointed out that group differences become important to the extent that our society bases policies on identification of people as members of groups rather than as individuals. Such policies are based in part on underlying assumptions about the distribution of abilities, and so these assumptions must be scrutinized.

Lee Cronbach indicated the need for a social philosophy that takes into consideration benefits other than efficiency and utilization of talent. Professor Thomson supported this suggestion: it must not be assumed that the only concern in connection with the development of social policy is whether or not a program will lead to maximal efficiency. Dr. Davis agreed enthusiastically: at this point he wished to suggest only that one should keep in mind the expectation from evolutionary theory of substantial variance in *all* traits that have a genetic component; not only intelligence, but also other talents and other socially valuable behavioral traits such as altruism, whose evolutionary origins had been discussed at length. At a later time the group will consider the possible implications of this diversity for public policy. For example, to anticipate a bit, one might conclude that in the interest of peace and harmony in society it is appropriate to have racial quotas; but if this is done it should be with an awareness that the result will not likely be congruent with that which would be obtained from a policy of equality of opportunity, for such congruence would be found only if abilities were randomly distributed.

Professor Mayr observed that throughout history, from the agricultural to the urban to the modern, highly technological society, the diversity of opportunities has been steadily increasing. The roles of primitive hunter-gatherers were mutually exchangeable, but since then specialization of occupations has been increasing constantly. This great diversity of opportunities is one of the values of human society, in that any individual, if he is properly helped by his education and by society, can find a particular niche for which he is particularly well suited. Past neglect of the fact of diversity has caused a great deal of unhappiness, when people were misplaced because the principle of equality, carried to the extreme, led to the assumption that everybody was identical in his abilities and capacities. Also, it is very interesting to observe how the prestige value of different occupations is constantly changing, and how it differs in different cultures. An interesting area for further study would be the relationships between the diversity of mankind, the diverse opportunities for jobs and professions, and the interactions between prestige values and the search for an optimal distribution of diversity in our society.

Professor DeVore pointed out one problem of any discussions of the degree of selection for different behavioral traits in widely separated populations: people realize that value judgments enter into such discussions almost immediately. To be sure, when one observes the amount of physical variance it is reasonable to conclude that behavioral traits vary as well. However, to admit that is to start down what in many people's minds is an inexorable path that leads to the presumption of significant differences in IQ between populations, which is not necessarily the case since intelligence is so polygenic and environmentally influenced. But even less significant behavioral traits have profound cultural and value orientations. Eventually politics becomes involved, because the group in power defines as positive values those traits which it possesses.

Philip Heymann found some difficulty with the idea of keeping in mind the fact that there may be significant racial differences. We live in a society where esteem, including self-esteem, is determined by the values of the entire society, and IQ, whatever it may measure, is taken to be a very important part of esteem and self-esteem. Hence, if we keep in mind that a certain group has genetically a population distribution with a lesser IQ, we are presupposing something that is very destructive of self-esteem, which makes it more difficult to deal appropriately on an individual level.

Dr. Morison suggested that one of the seminars be devoted to a direct approach to the problem of inequality. Many societies, and ours particularly, have hoped to reduce inequality in status and income, and it was generally assumed that if this could be done most problems would be resolved. However, as Christopher Jencks has pointed out, there is a great deal of variety of genetic background and education, and it is poorly correlated with status and income. That raises two questions: do we want to have the existing differences in status better correlated with some of the other features, or do we want to reduce or do away with such differences. If we want the latter, we have to approach the problem directly.

## HISTORY OF CONTROVERSIES OVER THE SOCIAL IMPLICATIONS OF EVOLUTION AND GENETICS

In a presentation dealing with the differing responses to Darwinism, Mark Adams first called attention to the *religious reaction.* He pointed out that Darwin did not publish his theory of evolution by natural selection for many years after completing the general scheme, probably in part because he expected religious opposition. When the work was finally published its conflict with natural theology did indeed create a strong negative reaction in England. In natural theology, the design in nature—the incredible adaptation of organisms, the fit of organic forms to the functions that they serve and to the conditions of life—was seen as proof of the existence of God. How else, it was reasoned, could such fine tuning occur but through the divine plan? This view created a kind of symbiotic relationship between science and religion whereby religion could not be threatened by the results of science because they were proof of the divine plan. To many of the religious Darwin seemed to be saying that while it is true that these fine adaptations occurred, some forms are more adapted than others, and it is precisely these differences that indicate a process in operation which governs evolution and which he called natural selection. Therefore, Darwinian theory was seen not only as a threat to views of creation, but also as undermining one of the most popular proofs of God's existence. Professor Trivers added that the conflict was due not only to the belief that design in living creatures was an important argument for the existence of God, but also to the belief in an additional design: a superplan in which the different species fit together in a harmonious universe. As is clear from Darwin's letters, he appreciated that his theory was fatal to this belief in a larger plan.

Professor Adams indicated that not very much is known about the differential response of different religious groups to Darwinism, or about the relation of their responses to their belief structures and to the theoretical predisposition of the community. However, there certainly were other negative reactions to Darwinism that have nothing to do with British natural theology. There is a great deal of diversity between the Protestant and Catholic response, and different Catholic communities responded in different ways at different times. Some religious structures probably denied that any scientific theory about the natural world could have any relevance for fundamental religious beliefs, while others felt threatened by apparent contradictions. Professor Mayr added that the situation is very complex in that the response to Darwinian theory was different from country to country; for example, in France evolution was hardly mentioned, not because of religion but because of the predisposition of the scientific community, while in Germany Darwin found more adherents of his theory of evolution than in Britain.

Professor Adams continued with a discussion of the interaction in the late nineteenth century between *Darwinism and social theory.* Darwinism meant different things to different people, and different parts of the theory appealed to different groups. In addition, Darwin is one of the least read of the late nineteenth century visionaries: most of what people know of Darwin is usually derived from Spencer, whose view of evolutionary development tended to be orthogenetic. In any case, for about 50 years after the publication of *The Origin of Species*, Darwinism, however defined, had a tremendously broad reception and influence among social theorists of all kinds, including sociologists, cultural and other anthropologists, ethnologists, linguists, social Darwinists, and the left-of-center political theorists (anarchists, Marxists, socialists).

Professor Adams cited three reasons for the significant impact of Darwinism on social theory. One was the widespread interest in natural history at the time. This term encompassed a wide range of subjects—geography, geology, mineralogy, climate, all aspects of flora and fauna, and very often the customs and behavior of human populations—and the natural historians looked to Darwinism as a theoretical contribution from which they could derive some insight about the relationships among these aspects of the natural world. Second, social science was developing at the same time, and the attempt to conduct these studies in a scientific way developed a keen interest in scientific theory. The third reason related to a concept of the interrelation of the sciences. Among late nineteenth century thinkers there was a view that the sciences had a kind of linear, orthogenetic scale. Many social theorists felt that biology was the science closest to their field, and they maintained that one should look at the development of biology to obtain clues as to what form the development of social theory should take. In cultural anthropology, sociology, and other fields Darwinism was seen as a theory that could provide a strong historical form of explanation for all processes of development. Other social scientists sought quite literally to explain social structure in terms of biological facts. Most of these social theories shared in common the orthogenetic trait; they perceived the application of Darwinian theory to their subject as involving the identification of a chain of development.

There are several reasons for the initial favorable response of left-of-center sociopolitical theorists to Darwinism. Probably the main reason is that they saw it as an undermining of religious authority, as rescuing an area of knowledge from the church. Also, to a certain extent they saw this model as another step in the process of acquiring knowledge that prepared the way for the social sciences, which in turn would lead eventually to the emergence of leftist political philosophy. Professor Mayr suggested an additional reason for the favorable response on the part of the anarchists: they were not only against the church but against any kind of authority, because it interfered with the freedom of man; they had an implicit belief in the basic goodness of man and felt that by removing all restrictions by authority the basic goodness of man would emerge. They viewed evolution by natural selection as a process that happens without outside authority, which coincided with their view of the ideal social order.

Professor Adams indicated that these political theorists also found the historical form of the theory very appealing; the notion of the historical development of man was one of their major concerns, and therefore they were attracted to a theory that did not invoke God or teleological views to account for historical development. However, they did not accept Darwinism without reservation. Their objection mainly took the form of deemphasizing intraspecific competition; they felt that the struggle in nature was overemphasized by Darwin, while cooperation and harmony were overlooked. They also tended to emphasize the inheritance of acquired characteristics. Although this partly relates to the state of knowledge in this area at that time, it also was due to the blurring of the line between biological and cultural or historical human evolution; if acquired characteristics could be inherited and passed on, one could use the same kind of theoretical framework to account for the evolution of societies and cultures.

Turning to more recent history, Professor Adams observed that one of the contributing factors to the development of Lysenkoism was the division within the scientific community, from 1900 to 1930, between what might roughly be called the naturalists and the experimentalists. The naturalists tended to support the Darwinian view of evolution, and were suspicious of Mendelian genetics as it was being formulated at that time, while the experimentalists tended to emphasize evolution by saltation. To a certain extent the artifacts of this conflict continued, even though a synthesis of naturalist and experimental approaches was developed in the 1920s. It is interesting to note that at no time has the scientific or the political line in the Soviet Union ever denied Darwinism; however, their acceptance took the form of maintaining the inheritance of acquired characteristics while deemphasizing natural selection and intraspecific competition, whereas the West reformulated its Darwinism on the basis of population genetics.

This difference became very prominent when Lysenko put forward his views on evolution. He contended that population geneticists were "fly-lovers and man-haters." Lysenkoists dismissed the study of the distribution of genes in populations of *Drosophila* because of its lack of social orientation; and Lysenko rejected the importance of intraspecific competition because in his view it would suggest that such competition occurred within human classes, when the real conflict is between different classes. Professor Adams observed that while Marxism may have been the ideological justification for Lysenko's theories, it is incorrect to conclude that all Marxists viewed Lysenkoism as being in agreement with Marxist philosophy. In fact, in the 1920s many dialectical materialists asserted that Lamarckism was idealistic and metaphysical and supported the study of orthodox genetics; they attempted to demonstrate that Lysenkoism was a wide aberration from Marxism.

Professor Adams concluded that the interactions between scientific discoveries and social theory are very complex and differ in different societies. Various groups accepted Darwinism on the basis of quite different interpretations of the theory. Evidently a typological point of view, abhorred in biology, may also interfere with the understanding of history; one has to look at the social goals and policies of various countries, as well as at their scientific views, in order to understand how ideas are interpreted. It is unlikely that any given idea will automatically lead to only one kind of social reaction; in different countries the same ideas can be used for different purposes.

Emphasizing the importance of the diversity of movements, especially in the biological sciences, Professor Mayr observed that there have been a number of attempts to apply Kuhn's theory of the nature of scientific revolutions to various developments in biology, but invariably they fail. In biology there are almost always two or three different paradigms coexisting simultaneously—e.g., epigenesis and preformation for 150 years, or Lamarckism and Darwinism for 120 years. One does not displace the other, but very often the best parts from each are eventually synthesized, as happened with evolutionary theory in the 1920s and 1930s. When this diversity interacts with contemporary movements in politics, sociology, religion, etc., very often there are unexpected influences on the development of science. Thus in Russia Darwinism was reinterpreted several times, depending on the dominant interpretation of Marxism.

Dr. Davis called attention to an interesting irony in the history of the conflicts of Darwinism with various currents of social thought. This theory has several major implications for our view of the nature of man. On the one side, the implications regarding man's origin were deeply offensive to many theologians; and the implications regarding change were troubling to members of the establishment who were concerned with maintaining the status quo. On the other side, the implications now emerging from the modern synthesis with genetics emphasize the existence of a great deal of genetic diversity within our species, thus challenging some liberal or radical beliefs. It is clear from the history of the earlier opposition to Darwinian theory that the facts of science will not disappear, no matter how much they may seem to conflict with prevalent political beliefs. Dr. Davis concluded that the achievement of social goals may be better facilitated by building on such knowledge rather than by denying it.

# Seminar II: Genetics
# (December 14-15, 1973)

**MAJOR PRESENTATIONS**

Bernard D. Davis: *Basic Principles of Genetics*

George Brewer: *Developmental Genetics*

Luigi Cavalli-Sforza: *Population Genetics*

*Co-Chairmen:* Luigi Cavalli-Sforza
Bernard D. Davis

## RECAPITULATION OF THE IMPLICATIONS OF EVOLUTION FOR HUMAN DIVERSITY

In a brief review of the material presented in the first seminar, Dr. Davis pointed out that it is most important for an analysis of human diversity to be as clear as possible on the nature of the evidence from evolutionary theory. This evidence shows that man is indeed a product of evolution, and the process by which this evolution takes place indicates the existence of wide genetic diversity with respect to all kinds of traits. It also suggests that subgroups within species that have been more or less reproductively isolated from other groups in the species for a considerable period of time will have accumulated significant differences in the frequencies of various genes.

A more controversial point is whether a knowledge of evolution leads to the expectation of diversity between individuals and between isolated subgroups of the population not only with respect to visible physical traits but also with respect to behavioral traits. Recent evolutionary history has been characterized by an extraordinary rate of change in behavioral traits, especially with respect to complex mental operations, which has far exceeded the rate of change in physical traits; given this history, it seems likely that conditions creating substantial visible differences in some populations would also create differences in the distribution of various behavioral capacities. In Dr. Davis's view, while the considerable reluctance to accept this conclusion is understandable, the evidence from evolutionary theory pointing in this direction should not be overlooked.

Noting that certain groups which can be separated on some dimensions would be mingled on a different set of dimensions, Daniel Bell asked whether there were any consistent sorting principles to define separations between subgroups in the population; how does one single out the more relevant dimensions for separations into statistically clustered groups? Professor Mayr replied that there are two basic principles relating to classification. One is that the more criteria that are used, the greater is the chance that the resulting information will provide a classification that is meaningful; the second, which is more controversial, is the idea that certain characteristics have more of an information content than others. If one looks at as many characteristics as possible, and determines which ones tend to go together and gives these extra weight, the result is something that is a reasonable indication of some sort of past history.

Dr. Cavalli-Sforza commented that because the available information from fossils is not sufficient to determine the evolutionary origin of human subgroups, one can only attempt to do a clustering on the basis of what one sees today; however, there is a continuity that makes every such clustering extremely difficult. He maintained that one can reconstruct the history only if one can rely on characters that have varied at random, which is not the case for many characters. Wherever there has been a good deal of nonrandom selection, it is extremely difficult to determine evolutionary origin from present-day data.

Professor Bell noted that in discussing diversity one is concerned with differential capacities to do different things; what is important is how one ranks the importance of the different kinds of capacities. The notion of natural selection implies selection in relation to the environment, which includes physical nature and the other species with which one lives. But in dealing with the problems of social selection one is interested in the kind of capacities people have to adapt to each other. For example, certain kinds of selective processes maximize people who have physical strength and endurance; certain others would maximize better conceptual thinking or the capacity to assess other people in a particularly sensitive way. These capacities may not be genetic at all but may be purely a social matter. A child who grew up in the ghetto is likely to be more perceptive about the motives of people than a middle class child in the suburbs, who lacks a certain kind of realism about the world. This is environmental; it has nothing to do with any genetic capacity.

Professor Trivers objected that the presence of obvious environmental influences on certain characters does not preclude the presence also of genetic components. Moreover, in *The Origin of Species* Darwin emphasized that other individuals within the species are a very strong component of selection; all of evolutionary social theory, which can be specified without referring to humans, is based on trying to determine how individuals were selected with reference to other members of their own species.

Since the period of separation of the major human races comprises only 1 to 5 percent of the two-million-year period of the high rate of evolution of behavioral traits, and since there was no detectable increase in cranial capacity during that time (although it increased two or three times in the two-million-year period), Professor Heymann asked how one can conclude that it is reasonable to expect differences in behavioral capacities among races because of evolution. Dr. Davis suggested that the rate of increase in brain size may not have dropped during the past 100,000 years: the available methods simply could not detect the difference that would have been produced in such a short interval. Moreover, since in the course of the last 100,000 years man developed very different patterns of social behavior—moving from hunting and gathering to agriculture to urbanization—one might even expect increased selection pressures for intelligence, and a few hundred generations would be more than enough for a good deal of selection to have taken place. It therefore seems unreasonable to expect that there has been selection for large differences in physical traits but for little in behavioral traits.

Dr. Davis further observed that the widely felt uneasiness about acknowledging possible differences between races may be due to the fact that discussion of racial differences is seen by many well-meaning people as racism. Two dogmas seem to polarize most discussions of this subject in this country today: the dogma that races are different and therefore that racial discrimination is justified; and, alternatively, the position that because discrimination must be opposed, it must be assumed that races are not different in their distribution of capacities. Dr. Davis suggested the possibility of evolving a third point of view: this would support objective analysis of the available evidence on such differences but would consider such information irrelevant to the decision to oppose racial discrimination; that decision should be based on moral considerations, rather than on assumptions about biological fact. In that connection, Professor Mayr emphasized that the phrase "races are different" should not be interpreted typologically; it does not mean that anybody of one race is different from anybody of another race, but rather that one finds a statistically significant difference between the means of two populations.

Professor DeVore called attention to the tendency of the anthropological tradition to suggest that natural selection, though critical for the human species up to the Pleistocene, was superceded at that point by cultural evolution; the strong implication is therefore that physical natural selection stopped. Since social scientists tend to rely on anthropologists for guidance in such matters, it is important to emphasize that such an implication is erroneous; physical natural selection has not stopped and in fact it may have been accelerating since the late Pleistocene. While there is no question that important social processes were added, that development did not negate ongoing natural selection. Professor DeVore stated that, in his view, if one finds significant morphological variations between two populations, one should expect to find a variation in behavioral predispositions. However, because behavior is under complex polygenic control and subject to substantial environmental variation, because our knowledge in this area is still very limited, and because the political implications of assumptions of populational differences are so far-reaching, caution should be exercised in discussing and especially in drawing any conclusions about such matters.

Lee Cronbach agreed with Professor DeVore that for the human species in general, and also within separated groups, genetic evolution has been continuing. However, in late Pleistocene there was a radical increase in the variety of cultural environments available, and this diversification would probably decrease markedly the predictive implications of a genotype for what would come out in behavior. The diversity of culture introduced a flood of new variance of environmental origins, both within and between groups, which would show up in the access to reproduction among other things. There may well have been a break in the link between the genotype and behavior, and it is not surprising that anthropologists would be concentrating on this cultural change almost to the point of implying that the other source of variance had stopped.

Professor DeVore referred to the evidence for variance in behavioral predisposition between Asian and European neonates. Dr. Cavalli-Sforza cautioned against interpreting those differences as necessarily genetic; if race influences behavior in the first 72 hours of life the difference is not necessarily genetically based. The child has had nine months of life in the hormonal environment of the mother, which in turn may have been influenced by the general environment. Professor DeVore expressed his willingness to believe that a hormonal difference might be involved. However, the kind of response that the mothers of a particular population had to the environment, and the hormones they released as a result of that response, have been under selection pressure.

Noting that an inherent limitation in genetic studies must be recognized, Dr. Davis maintained that one should not require absolute proof before drawing any conclusions from the available evidence. There are many areas of science where it is impossible to obtain rigorous evidence; and absolute proof, in a philosophical sense, does not exist in empirical science at all. In this case, if Oriental newborns are more placid than European, it is remotely conceivable that the mothers' hormones are responsible for the difference in behavior; and though there is no evidence for any large environmental influences on the mothers' hormones, such influences are also conceivable. However, it seems much more probable that there is a large genetic component. Dr. Cavalli-Sforza agreed that one may never obtain absolute proof on such questions, but he emphasized the necessity for being careful and rigorous, especially in this field.

Referring to a recent article that described the contradictory assertions of biologists about the effects of racial interbreeding, Professor Trivers also urged the avoidance of the tendency to assert positively what is only probable. He suggested that scientists discussing such issues have an obligation to state with rigor what alternative possibilities are still consistent with the data, even if their probabilities are weighted very differently. Dr. Davis agreed that one should acknowledge that a particular conclusion is not necessarily the only one consistent with the data, and that the focus should be on assessing the level of probability that various conclusions are valid. That does not mean, however, that one is justified in saying that certain evidence is worth nothing simply because it cannot lead to a definite conclusion with a high standard of rigor. Convergent evidence from many directions is usually the basis of growing confidence in a given theory in biology.

Professor Thomson expressed the view that if someone suggests that a certain body of data does not justify a certain conclusion, the person who says that should indicate what more proof he would require; if one does not think that the available evidence establishes a trait as genetic, he should indicate what further proof would be necessary. Professor Mayr noted that this is a very basic question of the philosophy and method of science. He pointed out that in science it is no longer considered necessary to "prove" something; if a well substantiated statement is not believed by somebody, he cannot discount it merely by saying that it has not been proved. Unless the evidence is very weak, it is up to the critic to try to refute it. Dr. Cavalli-Sforza indicated that because of the limitations of methods for analyzing polygenic inheritance (which will be discussed later) it is difficult to substantiate estimates of the genetic components with any precision.

## BASIC PRINCIPLES OF GENETICS

Turning to the topic of the present meeting, Dr. Davis indicated that this seminar would be devoted to a summary of the present knowledge of genetics, because this field throws additional light on the origins and the nature of genetic diversity in the human species. He began by describing some basic concepts of genetic theory. *Genes* are defined as functional units of inheritance that are located in a sequence on chromosomes; in terms of molecular genetics they are regions of a continuous, very long DNA molecule. When one says that an individual has a gene for a certain trait, he is referring to a particular form of a locus on a chromosome responsible for a given product. The gene may produce the normal product, or any of several different normal products (polymorphism), or a grossly defective product, or none.

The alternative forms of a gene, resulting from mutations and then inherited, are called *alleles*. The term "gene" is used confusingly, sometimes denoting a locus and sometimes an allele. In a diploid organism such as man, with paired chromosomes, an individual may have two identical or two different forms or alleles of a particular gene. Molecular genetics has shown that many genetic loci are occupied by a variety of alleles, each differing slightly in the sequence of the DNA and the protein it specifies. These differences may or may not result in detectable differences in function.

The property of *diploidy* (i.e., presence of a double set of chromosomes) permits a species to retain within it a great range of alleles that may not be expressed in one individual but may become expressed in another individual through the lottery that results in new recombinations. Generation after generation no two individuals are alike (except for identical twins), because of the reservoir of variations in genes that appear in new permutations and combinations in every throw of the dice.

*Homozygosity* refers to the presence of two identical alleles at a given locus, while the term *heterozygosity* refers to the presence of two different alleles. In a heterozygote one allele may be dominant and the other recessive, as in Mendel's original observations; but many genes exhibit codominance or partial dominance. *Genome* refers to the sum total of genes in any one individual; *gene pool* is the sum for the whole population.

The notion of phenotype vs. genotype is fundamental to genetics. *Phenotypes* are what is seen; they are the product of the interaction of the environment with the genome or *genotype* of an individual, which gives him a set of capacities that are modulated by the environment. For the bulk of the genes that are interesting in human beings, the gene determines a capacity to make something, while the amount of it that is made is influenced by environmental factors. It should be emphasized, however, that environmental influences have no effect on the genotype except through selection pressures on successive generations; that is, there is no inheritance of acquired characteristics.

Dr. Davis then proceeded to a discussion of *crosses*, pointing out the unusual coincidence that in Mendel's experiments with the sweet pea the number of characters that he described equalled the number of chromosomes, and every one segregated independently of the others. (He probably withheld aberrant data, but this would be forgiven today on the grounds of pioneering, rather than considered dishonest.) Subsequently it was found that there are genes for many different traits on the same chromosome, and they don't segregate independently. This led to a good deal of confusion until Morgan found that there is partial segregation of different traits whose genes lie on the same chromosome, and that the frequency with which they will segregate depends upon the distance between the genes. The farther apart two genes are on a chromosome the more chance there is of a recombination between them, and therefore the more frequently they segregate. If it is a long chromosome and they are towards its opposite ends, they will behave as if they are on separate chromosomes because the chances of recombination are so great.

Mendel dealt with easily segregated monogenic traits, for which mutation of one gene would result in a phenotypic change that could be scored cleanly. Morgan and his school, in their work on *Drosophila*, dealt with a much greater variety of monogenic traits, and they were able to map them in groups or clusters on chromosomes; but again they dealt only with those traits where one could score a phenotypic change and follow its distribution among progeny in terms of the distribution of single genes. However, the traits that are of most interest to the present inquiry are all very highly *polygenic traits*, which are far more difficult to analyze. For a time, many geneticists felt that while certain all-or-none aspects of inheritance were governed by Mendelian genes, other attributes, exhibiting continuous variation, may be governed by a very different form of inheritance—so-called blending inheritance. It took some time before it could be shown that such traits have fundamentally the same kinds of genic determinants; it is simply that more than one gene contributes to a given trait.

Considerable discussion was devoted to an effort to refine the definition of *trait*. Dr. Davis noted that in genetics it is a broad concept, referring to any component of the phenotype, and that almost any such aspect of an individual, physical or behavioral, has both genetic and environmental components. Professor Griliches felt that the definition of trait should be further limited in some way. For instance, the Chinese language is a component of the phenotype, and therefore a trait, but it has no genetic component. Dr. Davis replied that in connection with the knowledge of a language, the genetic component, differing among individuals, would be the distribution and effectiveness of the connections in the brain that provide the capacity to learn any language; what language is learned is determined by the specifics provided by the environment, and how well it is learned depends on both capacity and environmental input. Richard Herrnstein pointed out also that it is possible that a group of people speaking Chinese may vary in fluency for genetic reasons.

Professor Thomson also felt that there is a limited range of features that one would define as traits. For instance, it may be a feature of a person that he has just traveled on the subway, and yet presumably that is not a trait. Dr. Davis felt that such a property involves a whole set of traits; one should try to define trait in terms of those components of a person's behavior or physical properties that one can hope to correlate with differences in genes.

Professor Mayr observed that the word "trait" is probably more often used in taxonomy than in any other scientific enterprise; there trait is defined in a very vague way as some characteristic of individuals that helps to classify them. To classify a given species in terms of those who have just come out of the subway and those who have not is not a very informative classification, and all classifications are judged by their information content. One might define a trait as a characteristic of an individual or a group that gives some information that is useful for one's purpose. A sociologist, for instance, might be interested in income levels and might identify traits that are correlated with socioeconomic status. Talcott Parsons agreed that the term "trait" has a much more general sense, where the presumption of genetic influence may not be at issue at all. Dr. Davis agreed that a psychologist or a sociologist might be interested in a person's actual knowledge, and might call it a trait. The geneticist would not find that knowledge useful except insofar as it reflected a genetically conditioned trait: the person's capacity to learn.

Commenting that all behavioral traits are the sum total of learned items, David Layzer questioned the statement that a geneticist would not be interested in what an individual happens to have learned. Dr. Davis replied that the specific item of learning is not a trait of direct interest to geneticists insofar as the genetic component of behavioral traits relates to the capacity to learn. It is necessary to measure things that have been learned to determine the capacity to learn, but a specific learned item need not be considered a trait from a genetic point of view. However, Professor Layzer maintained that if a trait is to be defined as a component of the phenotype, as something that can be observed and measured, then the term "capacity" cannot possibly enter into the definition of a trait.

Gerald McClearn maintained that it is possible to measure a behavioral capacity. If one were to measure individuals at one point in time, subject them to some sort of environmental or educational experience, and measure them at another point and see how much change has occurred, that might be regardable as a measure of the capacity to learn, given that particular educational experience. In his view, a phenotype is anything that can be measured, anything to which one can apply some rule of measurement that will result in either a classification or a number that locates somebody on a scale. To ask whether it is genetic or environmental is meaningless because both genes and environment are necessary to maintain the organism that has that trait. The proper area of investigation is the heritability of the phenotypic variance observed in a population—i.e., what proportion is attributable to the genetic differences that exist among the individuals in the population, and what proportion is attributable to the environmental influences.

Professor Griliches objected to the term "capacity" because it implies that there are ceilings and limits. Professor Cronbach observed that while there is positive evidence of a capacity's being greater than zero, the evidence for capacity being absent is much more troublesome because of problems of motivation. Professor McClearn pointed out that operationally an analysis of capacity must be restricted to a particular defined context. Dr. Davis noted that some term is required to describe the situation in which individuals are exposed to a given learning environment and one progresses at one rate and another at a different rate. That is what is implied by capacity. Professor Trivers emphasized that because a word like "capacity" has a common language meaning, one must be careful to make certain kinds of distinctions in that an individual could have the capacity but not the will to develop that capacity, and he might end up with the same reaction curve as the person who is less capable.

Professor Griliches indicated that the word "capacity" has been used in different ways in this discussion. It has been used to affirm that some stimulus will in fact produce a response, that there is a nonzero response. It has been used to indicate the difference in reaction curves where the slope of one individual is higher than that of another. (In Professor Griliches' view, reaction curves relate to differences in ability rather than in capacity, or at least to a difference in a response rate to a stimulus.) Another interpretation, which is more difficult to accept, implies that for any individual that line approaches an asymptote; the level of the asymptote is defined as the capacity.

George Brewer observed that there is no problem with the term "capacity" as long as it remains a theoretical concept: the maximal possible gain in the optimal environment. Professor DeVore agreed that that is a very important point; whatever word is used, what has to be understood is that the concept relates to an ideal form. When one observes first generation Japanese children towering over the heads of their parents, one suspects that the children have more nearly approached the environmental nutritional ideal than their parents. It is important to keep in mind the notion that there is a range within which the genotype is conditioning, and not strictly determining, even basic morphological traits.

Dr. Davis observed that the terms "ideal" and "optimal" introduce another difficulty because their definition involves a value judgment. The optimal environment for height may not be optimal for other traits. If that should be true for morphological traits like height, one can expect even greater difficulties when dealing with behavioral traits. Perhaps the concept of capacity in behavioral genetics is best considered analogous to the concept of capacitance of a condenser in electricity: the ratio of current taken up in proportion to the voltage applied, and not the maximal amount of current that can be accumulated.

## MOLECULAR GENETICS

Continuing the discussion of basic principles of genetics, Dr. Davis reviewed some of the mechanisms on the molecular level. As described by Watson and Crick in 1953, the genetic material consists of a double chain of DNA held together by paired bases of four different kinds, the sequence of which determines the nature of the genetic activity in a cell. This discovery was then combined with Beadle's one-gene-one-enzyme model: successive sequences of a few hundred of the bases constituting the DNA each comprise a unit, a gene, that gives rise to a product (usually an enzyme) that is part of the working machinery of the cells.

The double helix as the molecular basis for the gene has a number of important implications. First, it accounts for the *stability* of DNA. A single-chain molecule could not persist, because the DNA in the bacterial cell, if unstrung, is a thousand times as long as the size of the cell itself. Anything that large would be sure to have frequent chemical breaks at various intervals if it were a single molecule. The fact that it is a double, intertwined molecule has two consequences for stability. First, DNA is in a sense a one-dimensional crystal, and crystals are much more stable than molecules in solution; with each strand leaning on the other, it is much more resistant to damage by the spontaneous thermal motion of the molecules, which tends to pull at bonds and occasionally to break them. Second, when a break in a strand occurs, enzymes in the cell will identify it and repair it immediately while the ends are still held together by the opposite strand. (However, whenever there is a double break the two DNA fragments separate, and the break cannot be repaired by the cell because ends do not find each other again. This is the kind of lethal damage that X-ray causes in cells.)

This double-stranded structure has implications for *replication*, the capacity of one gene to make two of itself. Each of these chains contains exactly the same information as the other; if the sequence of one is known, the other follows automatically because of the base pairing rule (e.g., hearts ⇌ diamonds; clubs ⇌ spades). A DNA molecule multiplies itself in the following manner: an enzyme (a polymerase) that acts like a snowplow splits the DNA strands; the linked bases of each strand are now free to attract their complementary bases (in the form of chemically reactive small molecules in the surrounding solution)—that is, as the two strands come apart, each one selects the molecules that will be strung together by the polymerizing enzyme to make its mate. Since there is only one possible correct choice for each of the bases, the resulting double strands will ordinarily be exact replicas of the original. Occasionally, as in any enzymatic reaction, a mistake is made and the wrong base is inserted; that substitution is one basis for *mutation.*

The structure of DNA also accounts for the *recombination* (crossing over) discovered by Morgan. In a diploid organism the chromosomes of each pair came each from one parent, and each sex cell then derives one chromosome from each pair. However, in the process of forming sex cells (meiosis) the paired chromosomes do not always segregate intact, but they often exchange corresponding regions with each other; hence part of a chromosome came from one member of the pair and part from the other. The molecular basis is that the paired chromosomes "breathe" or open up enough so that the bases in a region are no longer paired with each other but are exposed. If such a sequence of available bases on one chromosome recognizes a complementary available sequence on the other chromosome, those two strands will temporarily pair. That is, for a certain short region there will be a sticking together of one strand from one chromosome and its complement from the other chromosome. The two chromosomes are then cut and reunited by enzymatic action in such a way that these regions cross over, thus enormously increasing the variety produced by the segregation of meiosis. Chromosomes do not always pair perfectly. If the pairing is out of register by one base, after the cross-over there will be one chromosome with an extra base and one with one base too few. This mistake will ordinarily inactivate the gene, for it will make all the translation (see below) beyond that point incorrect. Such aberrant (illegitimate) crossing over is a frequent source of deleterious mutations.

In response to Professor Bell's question as to why mutations are not more common, Dr. Davis pointed out that there exist, at least in bacterial cells, *repair enzymes* that can recognize a region of DNA where the bases do not match, cut it out, and copy the original. In mutants of bacteria that are defective in those enzymes the mutation rate is one thousand times higher than usual. Professor Mayr added that in higher organisms another reason why mutations are not more common is that most of the new zygotes formed from defective gametes die so early that they are not detected; such elimination reduces the visible mutation rate. Dr. Cavalli-Sforza commented that there must have been selection for lower mutation rates because a complex organism that had too many mutations would not survive. Professor Trivers felt that it is important to bear in mind that the genetic system not only produces the genetic variability that natural selection works on but is itself a consequence of natural selection; all the complicated mechanisms that affect variability of offspring are themselves under the influence of natural selection. Different arguments have been advanced with regard to the conditions under which mutation rates would be decreased or increased, depending upon the extent to which it is advantageous to have more or less variance in offspring.

Professor Thomson asked why some mutations keep reappearing. Dr. Davis replied that every gene in a given organism has a certain characteristic frequency of mutation; while one cannot predict when a mutation will occur, he can tell the reproducible frequency of mutation in a given organism for a given gene. Hence if the population is large enough that mutation will reappear at certain intervals. Professor Thomson then inquired why certain mutations, like dwarfism, are more common than, say, extremely long ears. Dr. Davis indicated that the frequency of a given kind of visible abnormality in the population (mutant phenotype) depends not only on the mutation rate but on whether any of various mutations can cause the change (e.g., perhaps 100 different loci can give rise to deafness); on whether multiple genes are required (a polygenic trait, such as ear length); and especially on whether the trait depends on a recessive gene, in which case the population can maintain a large reservoir of phenotypically normal heterozygotes whose occasional mating will produce a mutant phenotype.

Continuing his presentation, Dr. Davis observed that the structure of DNA also accounts for another essential feature of evolution: the capacity for *expansion of the genome*, which can happen in the following way. It has already been noted that illegitimate crossing over, due to a slight shift in pairing, can cause mutations by resulting in insertion or deletion of a base. Similarly, illegitimate pairing occurs rarely between nonhomologous, distant regions of a pair of chromosomes, since by chance short regions of complementarity are bound to appear in any long sequences of four different bases, and these will cause brief, transient pairing, which will result in crossing over if the right enzymes come along soon enough. In such abnormal crossing over, long regions of the chromosome can be transposed, resulting in duplication of one or more genes (e.g., A B C D becomes A B B C D). Nature can produce variations in the superfluous B that will not harm the cell, because the original B is unchanged and still performs its usual function.

Dr. Davis then presented the evidence from molecular biology which confirms *evolutionary continuity*. The two strands of DNA of an organism can be separated by raising the temperature, yielding a solution of single strands that are all tangled up. Under the right conditions a large fraction of these strands will go back into register with their complementary partners and form double strands again. If one mixes strands of DNA of a man with a small amount of radioactively labelled DNA from a monkey to see how much of that radioactivity will find human DNA and form double strands again (which are easy to separate from single strands) he will find that this hybridization is a little less extensive than that of human + human DNA. There is less hybridization of human DNA with that of more distant mammals, even less with fish, and none with bacteria.

These results give an overall measure of the extent of complementarity between the two species. This molecular measure of the gradual accumulation of mutational differences in the sequence of DNA corresponds very well to the distances between species on the conventional taxonomic tree, providing the most direct proof of the reality of the evolutionary relationship between living organisms, and of the role of mutations in creating the variations on which evolution is built. In a parallel approach, the amino acid sequences in homologous proteins are compared—for example, in hemoglobin the chimpanzee does not differ from man, and the gorilla differs in one amino acid out of 146.

*Gene expression* involves transcription of the information in one of the DNA strands of a gene into the complementary RNA molecule (employing the same mechanism of base pairing as in gene replication). A few of these RNA molecules are part of the working machinery of the cell, but most are transient *messenger* RNA molecules; these are translated into the main elements of the working machinery, the proteins (mostly enzymes). The sequences of DNA are thus translated, via the genetic code, into corresponding sequences of twenty different amino acids in proteins; and this one-dimensional structural specificity is converted into three-dimensional specificity by the spontaneous folding of the protein chains, guided by the mutual attractions and repulsions of the various amino acids. The varied surfaces provided by this folding are complementary to other molecular surfaces and thus account for a great variety of specific interactions: the capacity of enzymes to bind specific compounds and catalyze chemical changes in them, the combinations of antibodies with antigens, and the aggregation of various proteins and various enzymatic products to form membranes, cells, organs, and, finally, whole organisms.

*Regulation of gene expression* is also increasingly understood in terms of the structures of DNA and proteins. Various genes are translated at different rates, in terms of both inherent capacities and environmental influences. In the regulation of capacity each gene (or a limited group of genes) has an adjacent region, the *promoter*, where transcription starts; and the affinity of the promoter for the transcribing enzyme limits the maximal rate of transcription. (This rate varies from one gene to another, and it can be modified by mutation.) For environmental regulation, a length of DNA between each promoter and its subject gene(s) serves as a gate, the *operator*, which lets the enzyme pass when open but will block transcription when combined with a *repressor* protein specific for that operator. This protein responds to variations in the concentration of a particular small molecule that can bind to the repressor and change its conformation. In one conformation the repressor can bind to and block the operator; in the alternative conformation it cannot. These special, flexible, regulatory proteins are the means by which genes sense the environment and alter their activity.

There is little doubt that comparable regulatory mechanisms, multiplied many times over and with very complex networks of intercommunication, must be present in animal cells and govern their interactions with the environment. The conclusion from genetics that individuals differ in their built-in capacity for a given function, above which an individual cannot be stimulated, finds a general explanation in this recent insight into how genes work at a molecular level. With a mutant bacterial strain having a poor promoter, which permits it to make only half as much of a given enzyme as most strains, there is nothing that can be done in the environment to make the cells make more of that enzyme relative to other enzymes. The only way is by a new mutation that results in a faster acting promoter. An organism can thus change its potential to carry out a certain enzymatic reaction by changing the amount of the enzyme, through mutation of regulatory genes; it can also make a more or a less active enzyme, through mutation of a structural gene.

Molecular genetics can also explain the concept of dominance, derived from classical (transmission) genetics. The two alleles in a heterozygote usually produce the two corresponding proteins. If the normal allele produces an effective enzyme and the mutant allele produces a corresponding defective one, the organism will usually suffer little if at all from a decreased level of the enzyme. Moreover, feedback regulatory mechanisms may even maintain a normal level, if a single gene has sufficient capacity. But if the product is a structural or a regulatory protein the abnormal protein may seriously impair function, even in the presence of its normal homologue.

Similarly, knowing that mutations affect the phenotype through changes in protein sequence and therefore in three-dimensional folding, and knowing that some amino acid substitutions have more effect on that folding than others, it is easy to see why various mutations produce a range of effects on an enzyme: from no detectable change, to decreased activity, or to greater sensitivity to various changes in the environment, and finally to complete loss of activity (the usual classical mutations). Incidentally, the possibilities of polymorphism are enormous. Thus because of the nature of the genetic code, all the possible one-base substitutions in the part of the DNA coding for one amino acid can cause it to be replaced by about six others; hence an average gene, coding for a protein 100 amino acids in length, could yield $6^{100}$ different products as a result of one-base substitutions.

Professor Mayr pointed out the importance of the mechanism for variable gene expression. Although the nucleus of every cell in a higher organism has exactly the same genes, cells with identical genes can produce a very different set of products because most of the genes are dormant in a cell at any one time. Only certain ones are activated, and they are the ones that produce the specific products that permit that cell to perform a specific function in the body.

## GENETIC DEFECTS IN MAN

Using the example of sickle cell anemia, Dr. Brewer illustrated the extensive changes that can result from a single mistake at the DNA level. Hemoglobin, which is composed of several chains each of 146 amino acids, is a typical protein: it is present in red blood cells and is responsible for the vital function of oxygen transport. In the mutation that produces sickling of the blood cells there is a substitution for the sixth amino acid. When the gene for sickle cell hemoglobin is present in homozygous state the individual has only that hemoglobin in his red cells. Because such cells aggregate and plug up capillaries, especially when the oxygen supply is decreased, the disease has the potential for affecting any part of the body, and there are many different manifestations—abdominal pain, stroke, heart problems, etc. This pattern illustrates the fact that one mutant gene can have many different effects. In an analysis of behavioral traits one should not exclude all the single-gene possibilities simply because the manifestations are complex, for though many genes contribute to the central nervous system, one gene may have many effects on it.

Dr. Brewer called attention to a social problem in connection with sickle cell anemia. Ten percent of American blacks are carriers of the sickle cell gene—that is, they are heterozygotes with one gene for normal hemoglobin and one gene for sickle cell hemoglobin, and they make about half of each type of hemoglobin (for the alleles act independently); they themselves are not ill. In an effort to try to reduce the birth of persons with the disease, sickle cell screening has recently been made mandatory in a number of states. However, such screening programs were administered in an unsophisticated way with no genetic counseling regarding the risk of producing the disease in progeny or the significance of the presence of the "defect." Moreover, there were cases of job discrimination against heterozygous carriers even though they have no significant disadvantage. Most of these laws have been repealed because of opposition from the black community, which originally favored this attention to the problem.

In response to an inquiry about the nature of genetic counseling, Dr. Brewer indicated that the prevailing philosophy is merely to call attention to the facts, the potential risk of having affected offspring if heterozygotes for sickle cell hemoglobin intermarry. Such information is not likely to affect marriage choice, but educated couples who are carriers may be convinced of the advisability of adopting; early identification of the disease and abortion is another possibility. Dr. Cavalli-Sforza commented that if specific advice were to be given, it is not that heterozygotes or even homozygotes for sickle cell anemia should not marry; the important thing is that if they choose to marry a normal, they would avoid entirely the chance of progeny affected by the disease.

Dr. Brewer then turned to a discussion of *chromosomal abnormalities*. In the body cell of a man there are 46 chromosomes (23 pairs) that can be distinguished under the microscope: 44 autosomes and the sex chromosomes (XX in females, XY in males). There are many chromosomal aberrations that can be identified—for instance, in chromosomal number the classic example is the occurrence of three of the small chromosome 21, which cause a disease called Down's syndrome (mongolism) involving a variety of abnormalities including mental retardation. This arises not through mutation or inheritance but through a process called nondisjunction, whereby chromosomes may stick together and during meiosis both of a pair may be dragged over to one pole, so that one gamete gets two doses of the chromosome, and another gamete gets none. The two doses, plus one from the other parent's gamete, yield a zygote with three. This trisomy causes harmful effects because in a complex organism the correct dosage of the genes has been carefully evolved and the excess dosage causes metabolic imbalances. Trisomy of the larger chromosomes is probably lethal because with the larger number of genes there will be even more imbalance. This disease presents another social problem, because its frequency increases markedly with maternal age. Women pregnant over the age of 40 have about a 2 percent chance of having a child born with mongolism. Since the cost of institutionalizing such an individual is about $100,000 per lifetime, the question arises as to whether diagnosis by prenatal techniques should be required of all women who are over a certain age, followed by abortion of affected fetuses, or whether an individual has a right to have whatever children she wants despite the consequences to society.

## DEVELOPMENTAL GENETICS

Developmental genetics is concerned with the mechanisms by which the fertilized egg grows into cells of all types, which have the same set of chromosomes but perform very different functions in the body. Dr. Brewer emphasized that although the systems of regulation in bacteria discussed earlier are well known, knowledge of such mechanisms of gene regulation in higher organisms, with their additional process of differentiation of a variety of stable cell types, is very limited. It is thought that the hormonal systems may be at least part of the messenger system involved in the process, but for the most part the mechanisms remain a mystery.

The area called developmental genetics, therefore, is more or less a descriptive field, where an embryo is studied during various stages of its development from the fertilized egg. There are specific genes that affect the morphology or the behavior of the developing embryo, and one can identify mutated or abnormal genes that interrupt some specific function: i.e., perhaps the gene operates improperly at some stage during development and thus interferes with the development of the brain. Although we do not know the mechanisms that make the parts of a developing organism come together at the proper time, timing is absolutely critical. As an illustration Dr. Brewer described how cleft palate may result from an error in the rate of growth.

Development is influenced, of course, by environmental as well as by genetic factors; the German measles virus is an example of an environmental agent that results in developmental abnormalities. The situation is exceedingly complex and involves various interactions; as Professor McClearn noted, a given environmental situation can have different consequences depending on the genotype of the individual.

Dr. Davis called attention to an important aspect of gene regulation in man which differs from that in bacteria. In the bacterial regulatory system there is a reversible system, e.g., when lactose is provided it affects a regulatory protein and the genes for using lactose are turned on; when it is taken away, they go off again. However, when an animal cell is committed to being a brain cell, there is some kind of autocatalytic system, about which nothing is known, that keeps it functioning as a brain cell. Dr. Brewer noted that there is reversibility in cell differentiation in plants; for example, with carrots a root cell can generate the whole plant.

## THE NERVOUS SYSTEM

Dr. Davis then continued with a brief discussion of the development of the nervous system. The brain is a giant computer in which there are about ten billion cells; each cell is estimated to have, on the average, a thousand connections with other cells. Impulses pass through these connections or synapses and influence, either in a stimulatory or inhibitory manner, the behavior of the next cell. These interactions determine neural functions.

There are several levels at which genes can be predicted to influence the formation and the function of the brain. First, genes must influence the actual pattern—the wiring diagram—of the brain. Although no one has any idea how many genes go into determining the pattern or how specifically it is determined, there must be very many genes that influence the development of the gross anatomical structure of the brain and the nature of its connections. Second, genes no doubt influence the nature of the switches; just as a mutation that changes one amino acid in a hemoglobin makes it function quite differently from another hemoglobin, some individuals will surely have some proteins in their synapses that will differ from those in another individual and will make the threshold for the firing of the switch, or its rate of recovery, somewhat different. Third, the behavior of the nervous system can be influenced by the composition of the fluid surrounding the cells, including variations in hormones, levels of calcium and other ions, and various drugs. These levels, and the response to their differences, must also have a genetic component.

Commenting that certain kinds of behavior have a greater tendency to be fixed while others are more variable and need more input from the environment for normal development, Professor Bell suggested that as the complexity of the organism is increased perhaps more environmental elements are involved. Professor Mayr felt that it would be difficult to draw such a conclusion with any certainty; for example, adaptation in bacteria is extremely environmentally determined.

Dr. Davis noted that there is now intensive effort, with mutants of simpler animals (*Drosophila*, worms, leeches), to define the relation between genes and the structural and functional features of the nervous system. As knowledge in this area increases, hereditary differences that are important for behavior will be identified, one hopes, in neurophysiological terms. However, at the present time the mechanisms building genes and behavioral differences are almost entirely a black box. Hence one's approach to this field resembles the epidemiological approach to an infectious disease for which one cannot isolate the organism. For example, hepatitis has been known for decades to behave in every respect like a viral infection, even though the virus was demonstrated only recently. In the same way, even though the genes that affect behavior cannot be counted, catalogued, or identified in terms of their specific actions, their existence and the general magnitude of their role can be inferred with assurance.

## POPULATION GENETICS

Because man is not an experimental animal and one cannot set up crosses, which is the standard genetical technique, one has to rely on the matings that occur—that is, on population data. That is why, Dr. Cavalli-Sforza explained, one has to study population genetics in order to understand human genetics. Even if this involves some oversimplification, population genetics usually deals with individual genes. Since genes are always in pairs in diploid organisms, and at one locus there may be several different alleles of a gene, it is considerably simpler to study the frequency of individual alleles than that of the possible pairs. Some information is lost in this process, but it can be recovered later.

**Computation of Gene Frequencies**

Evolution can be described as changes in frequencies of genes. In explaining how gene frequencies are computed Dr. Cavalli-Sforza used as a specific example the allele for sickle cell hemoglobin, which, as indicated earlier, differs in one base in the DNA and therefore in one amino acid in the hemoglobin. In a given population individuals may be of three types: those with both alleles that produce normal hemoglobin (AA); those with one normal and one sickle cell allele (AS); and those with two sickle cell alleles (SS). Suppose one finds in a population of Africans the following frequencies of these genotypes: AA—81 percent; AS—18 percent; SS—1 percent. To determine gene frequencies one has to count how many genes in the population are A or S. Assuming that the indicated population is composed of 100 individuals, the number of A genes is 81 × 2 from the AA individuals plus 18 × 1 from the AS individuals, totalling 180. This total must be divided by the total number of genes being considered (100 individuals with two genes each = 200); therefore, the frequency of the A gene in this population is 90 percent, and the frequence of the S gene is 10 percent. Under conditions of random mating, one can compute genotype frequencies given the gene frequencies by means of a very simple method called the Hardy-Weinberg theorem: the frequency of genotype AA is the square of the gene frequency for the A allele; of AS the product of the two gene frequencies times 2; and of SS the square of the gene frequency for the S allele ($p^2 : 2pq : q^2$).

Professor Griliches pointed out that for many interesting traits random mating is not a reasonable assumption. Dr. Cavalli-Sforza agreed that for many behavioral traits it is not random, in most cases because individuals choose each other by social class, for which these traits have a stratification. But there is completely random mating for almost all the traits that are called Mendelian in the sense that they follow exactly Mendel's rules—that is, those that are strictly genetic with essentially no environmental effect.

In answer to Professor Thomson's question as to what determines a gene to be an allele of another rather than a different gene, Dr. Cavalli-Sforza suggested an analogy whereby a gene might be seen as a word on a page of which several copies were made; an allele of that gene would be the word in one of the copies in which one letter is changed: the word is on the same position on every copy of the page, as the allele is at the same locus on the chromosome, but it is copied in slightly altered form. Professor Mayr added that at the next locus there would be a totally different functional unit—one might relate to the nervous system and the next to the blood; because of its difference in function, one identifies that as a totally different gene.

### Changes in Gene Frequencies: Mutation Pressure

Gene (and genotype) frequencies remain constant from one generation to the next unless they are affected by any one of three factors: mutation, selection, and drift. There may be, for example, an accumulation of S types in a population because of mutation; that is, simply because A tends to mutate to S. It is also possible that S might mutate back to A. However, mutation rates are very low, and in order to have substantial changes due to mutation alone, many generations would be necessary—approximately the inverse of the mutation rate. Thus the frequency with which A may mutate to S in one generation may be on the order of one in a million, and if this were the only mechanism it would take about a million generations for a population to change from mostly A to S, which is a long time even on an evolutionary scale.

### Selection

Selection is a much faster process. When an advantageous mutation arises in a population the advantaged individual will tend to leave more progeny than those who do not carry the mutant allele; and if the advantage remains more or less constant over generations, that gene will increase in frequency. The degree of advantage is defined by a quantity called the fitness or selection coefficient (the two are easily obtainable one from the other), which relates to the relative number of children produced by each of the genotypes.

In a population with the sickle cell allele the SS individuals usually die before they reproduce, and on average they leave only 20 percent as many progeny as the others—that is, their fitness is 0.2. In this country the fitness of the AS heterozygote is probably the same as that of AA. However, the heterozygote is believed to have an advantage relative to normal, AA individuals with respect to malaria; hence under malarial conditions the AS has the highest fitness (1.0), while the AA individuals have a fitness of about .9 or .85, about 10 or 15 percent lower. That is, in a malarial environment there is a 10 percent disadvantage of the AA type, and an 80 percent disadvantage of the SS type, with respect to the AS type.

Referring to data relating to various rates of increase in gene frequencies because of selection, Dr. Cavalli-Sforza indicated that with a selection coefficient of .1, if one starts from a frequency of 1 in 100,000, it will take on the order of 100 generations for the gene to reach a frequency of about 20 or 30 percent. If the frequency of the mutant gene reaches 100 percent the other allele is eliminated. That would not happen with the sickle cell allele, because the homozygote SS is at a serious disadvantage.

One can visualize the evolutionary process as follows: in one (or perhaps several) parts of the world the mutant arises in a malarial environment. In the beginning there are only heterozygotes; homozygotes will be extremely rare because, as indicated earlier, the frequency of the homozygotes is the square of the gene frequency. As time passes and the gene frequency increases, homozygotes start to appear, and since they will usually die, that will put a brake on the growth of the S gene frequency until eventually it will come to an equilibrium. Equilibrium is dictated by the relative disadvantages or fitnesses of the two homozygotes. This is an example of equilibrium due to selection in favor of heterozygotes.

Another type of equilibrium is that which can be established between mutation and selection. This would occur in the case of a completely deleterious mutation, a genetic disease in which there was no known advantage in the heterozygote. Thus phenylketonuria (PKU), a recessive disease, causes such severe mental retardation that individuals are essentially barred from reproduction, so that their effective Darwinian fitness is zero. The allele is produced at a constant frequency by mutation, but whenever it occurs in homozygous form those copies are eliminated because such individuals do not reproduce. Therefore, mutation produces the gene and selection eliminates it. Whether such a deleterious gene is recessive or dominant, it will accumulate in the population until the frequency is such that the elimination rate is equal to the production rate.

Dr. Morison suggested that one cannot completely rule out the possibility of heterozygote advantage in PKU. Dr. Cavalli-Sforza agreed that it is always possible, especially because of its high frequency (one in 20,000 among Caucasians). In fact, it has been postulated that there is a heterozygote advantage for another recessive disease, Tay-Sachs, a mental deficiency present in a high rate among Ashkenazi Jews. It has been suggested that this gene may relate to resistance to the TB bacillus. One should then also assume that conditions in ghettos of Ashkenazi Jews led to a high rate of tuberculosis. Since the evidence is vague, that is a very fragile hypothesis at the present time.

Dr. Cavalli-Sforza called attention to figures relating to selection relaxation for recessive traits kept in balance by heterozygote advantage to indicate what can be expected for the sickle cell gene in the future. With no heterozygote advantage, which is essentially the situation today for black Americans, after one generation the change is essentially a trivial one; after ten generations (about 300 years) some change is expected—heterozygotes might have decreased from an original value of perhaps 18 to about 10 percent, and homozygotes from 1 to ¼ percent—thus one might expect some difference in frequency between populations of blacks who have been in America (that is, in a nonmalarial environment) for two or three centuries, and those who remained in Africa. However, it takes a hundred generations for the disease to disappear under the effect of the natural selection against the homozygote. Dr. Cavalli-Sforza emphasized that changes in gene frequencies are always very slow.

Professor Bell asked whether it is possible in principle to have changes in all human genes. Dr. Cavalli-Sforza replied that in principle every gene can be changed but that many (such as those for behavioral traits) are part of a polygenic system. To predict the fate of such mutations one would have to know their interactions with the other genes of the system, as well as the selective forces involved. Dr. Davis added that mutations in many genes cannot be seen because they make the embryo nonviable.

### Drift

Another process that can change gene frequencies is drift, a term used to describe the fact that gene frequencies can change within a population in a random fashion. This is entirely a statistical phenomenon: every generation is a sample of genes from the previous generation, and a sample is always of finite size; sampling is responsible for chance fluctuations that will accumulate over generations. For example, suppose when the American Indians came to the American continent through the Bering Strait there were only ten individuals, all of whom had the gene for type O blood; that would explain why American Indians are all blood group O. However, other explanations of a selective nature are also possible. This is called the *founder effect* and is essentially an example of drift. The effect of drift is inversely proportional to population size: in a population of 100 individuals the population may change very drastically in 100 generations; in a population of one million individuals it will take a million generations to see a similar amount of change due to drift. Drift will occur despite the size of the population, but it is clear that the smaller the population the larger the likelihood of sampling errors.

There is a tendency on the part of some geneticists to believe that a large amount of the differences that one sees, for example, between the hemoglobin of a horse and the hemoglobin of a man is due to drift. This question is of some relevance to an analysis of the differences existing among individuals and what they may mean. When one says the difference is due to drift, one essentially means that it is not due to selection, and that the genes involved are selectively neutral—in other words, that one allele has no advantage over another. This may be true for some genes, but it is difficult to believe that it is true for very many, since selection is the only mechanism that can determine adaptation. In Dr. Cavalli-Sforza's view, it is likely that a fraction of the genes involved in molecular evolution are neutral, but it is difficult today to estimate the amount with any degree of probability. Professor Mayr felt that apparently neutral alleles may merely be ones whose selective advantage has not yet been identified.

Professor Feldstein observed that drift could operate simultaneously with selection. Dr. Cavalli-Sforza emphasized that there is always drift. Professor Griliches indicated that if one assumes that historically most of the human mating populations were relatively small, the role for drift would be much larger. Dr. Cavalli-Sforza agreed that it probably was larger, but he pointed out that there are some misunderstandings regarding the human populations in the Paleolithic: the density might have been low but the total numbers were probably relatively high. The probability of a neutral allele's being fixed is equal to its frequency: if its frequency is 10 percent, there is a 10 percent probability of its being fixed in the limit and 90 percent probability of its being eliminated.

#### Distribution of Polymorphic Proteins

Professor Herrnstein asked whether there is any evidence regarding what fraction of loci have only one allele. In reply, Dr. Cavalli-Sforza called attention to an important development in 1966, when independently Harris and Lewontin used electrophoresis (movement in an electrical field) to systematically examine blood proteins in order to determine how many showed differences, and with what distributions. By this process (which was first used by Pauling to show genetic differences in A and S hemoglobin), if a gene is represented by different alleles the corresponding proteins are likely to differ in electrophoretic mobility. Harris and Lewontin found differences in one-third of the proteins tested. This is a tremendous variation: extrapolating to all other genes (estimated at a minimum of 50,000), and with two alleles at one-third of the loci and with three genotypes for each pair of alleles, this would result in $3^{17,000}$ possible genotypes. In fact, the variation may be even greater, because electrophoresis misses a certain number of differences.

Professor Mayr emphasized the value of this technique: it permits determining the genetic composition of an individual from the molecular phenotype, eliminating the necessity for crosses. He also called attention to the historical importance of this research. The classical geneticists thought of individuals as being homozygous at most loci with an occasional mutation. Some people of the Muller school calculated earlier that no individual is likely to be polymorphically heterozygous at more than seven or eight loci, while other geneticists believed that there was great genetic variability in populations. Until this research it was not possible to determine with any degree of confidence which view was most probably valid. The continuing research, showing that very few species of animals have a low allelic variability, confirms the enormous variability of most populations at most loci. Dr. Brewer observed that this evidence has forced those who originally maintained that there was low genetic variability in the population to change their argument to one of low selective significance: they now contend that there are very few loci whose variations influence selection—the rest are neutral.

Dr. Cavalli-Sforza presented some statistical data relating to differences between individuals and between populations in 24 loci, chosen from those known to have at least two alleles in at least one population. When two individuals were both northern Europeans they differed on the average in 33 percent of these genes; an essentially similar figure was found for two central Africans and for two Orientals. The difference between two individuals taken from any two of the subgroups averaged a little higher, about 40 percent. This result suggests that genetic distances between populations of even the most different racial groups are fairly small.

Richard Freeman felt that the difference between 30 and 40 percent could be significant and asked how one evaluates whether such a difference is large or small. Dr. Cavalli-Sforza replied that if a European differs from an African in 40 percent of his genes and differs from another European in 30 percent, the difference due to race is only 10 percent, which is small with respect to the average difference between any two individuals taken at random.

Dr. Davis suggested that it may be an oversimplification to conclude from this kind of evidence that the differences among races are small compared with the differences within races. The question of whether the differences between large subgroups of the human species are large or small is too broad to have much meaning; the question always has to be asked, difference with respect to what? Differences among races would be large for genes related to those physical traits, or to resistance to infections, that have been highly selected for. Hence the distribution of blood proteins for which there is no evidence of much selection pressure tells one little about genes affecting traits for which there has been a good deal of selection pressure. We simply do not know which is the better model for various behavioral traits.

Dr. Cavalli-Sforza replied that one cannot yet say whether selection pressures have acted on more than very few of these genes. But in any case the question of the overall amount of differences among races is complicated by the fact that while the differences one sees—in skin color, facial features, body build, etc.—are large, there are good reasons to believe that these are not a random sample of genes. These genes have one common quality: they are adaptive responses to climatic environment. The major differences among the various types of habitats to which man has been exposed relate to climate, and regulation of adaptation to climate must occur through the interface between the body and the environment, because that is where the contact between the external environment and the internal environment takes place. Therefore, the sample of differences among races that one observes is likely to be a biased sample due to climatic adaptation. With a random sample the differences are far less.

Professor Griliches observed that there is also an implicit bias in saying that all genes are equally interesting. If one were to utilize a weighting scheme to indicate that some genes are more important than others, the differences between populational subgroups with respect to the important genes may be quite different than with those of less importance. Dr. Cavalli-Sforza maintained that because one knows nothing about what weights to give, it is preferable to deal with an unweighted sample. Observing that there have been continuing arguments among taxonomists as to whether characteristics by which one classifies groups should be given equal weight, Professor Mayr stated that no one now accepts the equal weight assumption. To be sure, assigning different weights is extremely difficult, but there is no argument that if equal weights are given the groupings are usually rather meaningless. Genes relating to adaptation to the environment will be very different in different races, while those relating to heterotic balance might be the same in different races. One would have to subtract the latter genes from the total in order to be left with the ones that were meaningful for comparison of different populations. Dr. Cavalli-Sforza noted that although there are some loci that have very high percentages of probability of differences and others very low, the standard deviations of the figures indicated are fairly low. It is interesting to note that the average difference between individuals within the same village is not much smaller than that between an Englishman and a Frenchman. The difference tends to increase very slowly with increasing geographic distance between populations, supporting the hypothesis that the separation among races is a comparatively recent phenomenon.

In connection with the issue of the amount and the meaning of genetic differences among populations, Dr. Davis suggested that a basic question is the degree of relevance of the observed biochemical traits as a model for other traits. For the former the variance within the defined races averaged about two-thirds the total variance; but it is not clear how safely one can extrapolate from this pattern to any broad generalizations about the differences among populations in other socially more significant traits that may have been subject to greater differences in selection pressures.

Professor Mayr called attention to an intuitive element in defining differences among groups: the man in the street would probably say the difference between Europeans and Africans is ten times as great as that within the European population. However, the data presented indicate that the differences are much smaller, and it might be interesting to speculate on this seeming discrepancy. Dr. Morison felt that no real discrepancy exists, that a difference that looks mathematically small can be very significant.

Dr. Davis noted in this connection that the difference among races for certain biochemical traits—one-third or one-half the total variation—would be socially significant if it applied to the genes for intelligence; it would coincide with the Jensen position that there is probably a large genetic component in the differences among races, as well as in the variance within any race. Dr. Cavalli-Sforza indicated that one has to make another assumption in order to support the position that IQ differences among races are likely to involve genetic differences: one has to assume that the genes for high IQ are systematically more advantageous in one population than in another. Otherwise the expected difference between two populations in genotypic IQ would be zero.

Professor Herrnstein questioned the value of a sample of 70 loci out of a total of between 50,000 and one million. Dr. Cavalli-Sforza maintained that if it is a random sample with respect to differences that exist, the result is meaningful. However, Professor Herrnstein pointed out that the sample consists of the 70 loci that can be measured and in that respect it is not random; he asked whether one could in principle subject any gene locus to the same measurement techniques. Dr. Cavalli-Sforez replied that this would not be possible with present techniques, but that does not indicate any bias.

Commenting that the two-thirds of the loci that show no difference must have been subject to a different kind of selection for homogeneity than the one-third that do show differences (or other genes that show much larger differences, as for color or sickling), Dr. Davis questioned whether one can extrapolate the degree of difference indicated by these figures to other loci, especially those concerned with behavior. Dr. Cavalli-Sforza commented that one might well expect a larger or a smaller difference with regard to behavioral traits, but there is no reason to presume that it would necessarily be larger. Agreeing that one cannot evaluate whether the sample is producing too large or too small a difference, Professor Herrnstein emphasized that his point was that the sampling procedure does not conform to the statistical model of pulling balls from an urn because in this case some of the balls will be very sticky; if the method cannot extract any gene from the urn it does not yield a random sample. Dr. Cavalli-Sforza maintained that it is random in the sense that it extracts from an urn to which there is access, the urn of serum proteins.

Noting that morphological differences between populations are large while the protein differences are relatively small, Professor Feldstein suggested that it is difficult to deduce anything with respect to behavioral differences since one does not know which degree of difference is relevant for behavioral traits. Dr. Cavalli-Sforza maintained that the sample of differences that is visible—that is, the morphological differences—is likely to be biased in favor of differences because of selection for climatic variation.

Observing that the indicated data relate to whether or not two individuals have exactly the same protein, Professor Griliches asked whether there is a measurement to indicate whether they differ in a greater or lesser degree—that is, by more than one unit. Also, he pointed out that the underlying assumption of this method is that alleles of the same locus all produce a particular kind of protein or something very similar to it; he inquired whether it is conceivable that the gene in one format would produce a particular protein and another would produce an entirely different protein that would not be recognized as a difference. Dr. Davis replied that if measurements are based on enzymatic activity an inactivated enzyme would not be detected, even though it is a product of an allele of the locus in question. Dr. Cavalli-Sforza added that there is no systematic bias due to that aspect. However, with regard to the question of the degree of difference between different forms of a protein, he indicated that electrophoretic mobility, while measurable, is largely unrelated to the function of proteins, so that it does not provide much information about how they are going to work even if it is very different in different cases. It is a physical property of the molecule, which has a very low correlation with the biological properties.

Professor Mayr observed that some amino acid replacements make very little difference, which leads the neutralist school to declare that such differences are of no selective significance. On the other hand, many known hemoglobin mutations are extremely deleterious. By looking at the amino acid replacements and knowing the particular protein, one can determine whether an exchange is more drastic or less drastic. But the differences observed in populations as polymorphisms are, with few exceptions, the ones that are close to neutrality, because if they were more drastic the favorable ones would be immediately taken over and established by natural selection and the deleterious ones would be eliminated. Dr. Davis added that since the one-step mutant in a given locus is rare, the multistep mutant—the product of two rare events—is not frequent enough to be significant. One usually finds multistep mutants only when one compares homologous proteins from different species.

Professor Layzer asked whether a small number of superficial differences entail a large number of genetic differences to accommodate them in a consistent genetic scheme. Dr. Cavalli-Sforza replied that such a conclusion cannot be excluded. Professor Mayr added that recent research indicates that genes are very often tied together into functional complexes, and therefore if there are allele replacements in some of the loci, this might require allele replacements at other loci that are not directly involved in the observed character. Dr. Cavalli-Sforza described a phenomenon in artificial selection that is fairly well documented called *correlated response*: whenever selection occurs other apparently unconnected things happen, either because they are related physiologically or because they are related genetically through linkage of genes. One trait that has been studied intensively in *Drosophila* is the number of bristles on a certain part of the body. If one selects for individuals that have the highest number of bristles on a certain segment of the abdomen one may also find other changes. If these relate to hair increase in other parts of the body it is likely to be physiological—what one is selecting for is a gene that increases or decreases hair everywhere. If one finds that the thorax has been affected in some way that may also be physiological, since not enough is known about the physiology to dismiss the possibility; but it is more likely to be selection due to linkage. This kind of selection is restricted to the region around the gene or genes for which one is selecting, and it may involve undesired changes—for instance, if one selects for a big bull he may also select for a bull that has heart trouble. In artificial selection one correlated response that is almost always observed is a decrease in fertility. Professor Herrnstein suggested that if the decrease in fertility occurs in selecting for a variety of other characteristics, that correlation would not be due to linkage. However, Dr. Cavalli-Sforza felt that it could very well be due to linkage because fertility is so complex: a great number of genes contribute to it, and so whatever trait one is selecting for is likely to involve a locus close to some of these genes.

Professor Layzer asked whether this could be a pleiotropic effect, that when one selects for a gene for pigmentation that could directly change other characters and therefore give rise to a different optimal kind of balanced genetic system. Dr. Cavalli-Sforza agreed with such a possibility but pointed out that the strongest effect is always for the trait for which one is selecting; all the other effects are likely to be secondary. Dr. Davis expressed the view that this is probably not a very important phenomenon in dealing with natural populations which have been stabilized and adapted to different climates and other selection pressures, because if genes for undesirable traits were linked and carried on with desirable ones, the undesirable ones would eventually have been eliminated.

## ANALYSIS OF POLYGENIC INHERITANCE

Dr. Cavalli-Sforza then turned to an explanation of the methodology for analyzing polygenic inheritance. Galton, one of the first to be concerned with quantitative traits, was responsible for some interesting advances in this area. In attempting to measure the similarity between relatives as a measure of the intensity of the heredity, he devised a quantity which is very similar to the correlation coefficient used today. Galton observed that the value of the correlation coefficient between a parent and child tends to equal one half; the regression of the child on the parent is also about one half.

Professor Layzer asked if the correlation of one half is due entirely to the random mating. Dr. Cavalli-Sforza pointed out that for many traits in man the observed correlation is a half, not because there is random mating and absence of environmental effects, but because there are several factors which compensate for one another. Professor Layzer felt that parent-child correlation and regression to the mean are two separate factors which need not go together; Dr. Cavalli-Sforza said they do in fact coincide unless there is a difference in the variance of parents and children. Professor Herrnstein emphasized that the correlation coefficient and the regression coefficient are linked together logically; it is neither a genetic nor an environmental fact, but a statistical fact.

Galton suggested a theory to explain correlations, called blending inheritance. This theory postulated that the contribution of the two parents is mixed so that there are no longer distinct units contributed by the father or the mother. However, in Mendelian inheritance the contributions of the father and of the mother are passed more or less independently to the future progeny. The general objection to the theory of blending inheritance was made by Fisher: if generation after generation the child is the mean of its parents this must have the effect of reducing the variation, since a mean has less variation than the individual values; and after a certain number of generations the genetic variation should be reduced to zero, apart from fresh mutation. Since it is not, under blending inheritance one would have to postulate a much higher mutation rate than is realistic.

Dr. Cavalli-Sforza cited several important contributions to the analysis of polygenic inheritance. Some basic experiments relating to environmental variation were made by Johanssen. After measuring beans to find the distribution of their weights, Johanssen grew them and self-fertilized the plants; he found that there was a tendency for the mean of the progeny of each bean to be very similar to the weight of that bean. He then self-fertilized these for a certain number of generations and created pure lines—lines in which individuals have a very high degree of genetic similarity. He found that these pure lines each had a variation in weight less than the original one; also, if he selected individuals which were higher or lower in weight within a pure line, their progeny was not of average weight equal to the parents but reproduced that of the line. In other words, artificial selection had no effect on pure lines, and the population reverted entirely to the parental mean.

Dr. Cavalli-Sforza emphasized that since these experiments were performed in as homogeneous an environment as possible, the variation remaining in pure lines can be considered as due to "environmental" variation that the experimenter cannot control. This is very important in its application to human considerations, because unlike a breeder, who can control the environment, those concerned with human variation have to take the data as they are. At best they can do what is called statistical control—measure such things as income and socioeconomic status and correct their effects on the trait examined, hoping that they reflect the parameters of the environment that matter.

Dr. Morison suggested that it would be more correct to refer to this as unaccounted-for variance; it is not genetic but it is not necessarily environmental. Dr. Cavalli-Sforza agreed that the term "environmental" is a misnomer because there could be some developmental chance processes that one cannot control in any way that produce variation. Professor McClearn felt that if environment involves everything that is outside the DNA code, such developmental accidents can be regarded as part of the internal environment. Dr.

Cavalli-Sforza suggested that in some cases it might be difficult to attribute the variation to the internal environment; it may be strictly indeterministic. For example, it may be that during development there are phenomena determined by a small number of molecules of which in one case five go to the left and four to the right. It would be difficult to call this environmental—it is something produced by the organism and the indeterminism is tied to the nature of the process.

Professor Griliches suggested that it might be called residual variance. Dr. Cavalli-Sforza agreed that "residual" is a better term. But one should also remember that even with considerable inbreeding the residual variation may have a genetic component. There is no inbreeding process that eliminates all the genetic variation; what remains is usually very difficult to eradicate, being perhaps due to gene complexes with strong heterozygote advantage. Professor Thomson asked whether mutation might be the explanation for the remaining variation. Dr. Davis pointed out that mutations are too rare, and Professor Herrnstein added that the variation does not breed true, as it would if it were due to a mutation.

Another important development took place in 1916 when East published the results of data relating to the inheritance of quantitative traits in plants, for instance, corolla length in the tobacco plant, *Nicotiana longiflora*. These experiments were among the first to show the existence of what are now called polygenes—that is, a number of genes that each have a small, additive effect on a trait. In this experiment two lines were crossed, one with a large corolla and another with a small corolla (this difference was genetic: it bred true). The F1 was intermediate, but since it was composed of individuals as homogeneous as the parents (though heterozygous for many loci), the individual variation was essentially the same as that of the parents. In the F2, however, the genes segregated at random, which yielded additional variation. This variation can be used to estimate the number of genes involved in the difference between two pure lines. In fact, the F2 variance (actually, its excess over that of the F1 or the parental ones) is smaller the larger the number of genes involved in the difference between the two parental lines: for the larger this number the more the average of the contributions of different genes, in the products of segregation in the F2, will yield a similar phenotype.

In 1918 there appeared a very important paper by R.A. Fisher called "On the Correlation Between Relatives on the Supposition of Mendelian Inheritance." This provided the theoretical basis for interpreting in terms of polygenes the correlations that Galton had interpreted on the theory of blending inheritance. Correlations between relatives, on the assumption that the inheritance of continuous traits is due to Mendelian characters, turn out to be exactly the same whatever the number of genes involved, on the simplifying assumption that the genes interact with each other in an additive way.

Fisher's paper did not have a great deal of impact until later. In 1948 Mather published an explanation of the theory, adding to it in various ways, among them the application to many examples in plants and *Drosophila.* In his effort to bring these data to the attention of geneticists, Mather made one hypothesis: he thought that polygenes were different in nature from major genes, that there are genes that determine only small effects on traits, and that major genes which determine large effects are quite different. While there may be some interest in Mather's hypothesis, Penrose pointed out that a major gene can be a minor gene on other occasions—for example, a major gene for PKU can be a minor gene for head size or hair color. It is possible, however, that polygenes might often be regulatory genes. Dr. Davis suggested that it also seems possible, with modern recognition of widespread polymorphism, that polygenes are not always genes at different loci but may be alleles with small gradations of difference in their activity.

The experimental breeder has a great advantage over the human geneticist: he can control the environment to a point where he can detect subtle genetic differences between strains, and he can make his strains relatively pure. Skin color is probably the only case in human genetics where one can use the pure lines approach to an analysis of inheritance of a polygenic trait, because natural selection has separated lines with high and low pigment levels in the skin. Neither line is really pure, for there is residual genetic variation in the pigmentation of both blacks and whites; but at least there is a large nonoverlapping difference in the average skin color. Observations have been made on the distribution of skin color, which can be measured by reflectance spectrophotometry fairly accurately. At all values Caucasians have a higher reflectance than blacks, and the back-crosses and the F1s tend to be intermediate roughly at the position where they would be if the genes were fully additive. In order to look at the number of genes one has to look at variance values for crosses of hybrids. These are in agreement with the existence of four distinct color gene differences between blacks and whites. A four-gene difference is a large difference, but it is reasonable to assume that there has been selection for black skin color in the tropics and for white in the temperate and more northern regions, so it is not unexpected. It would not take a very long time to build up this difference, with sufficient selection.

In the analysis of polygenic inheritance in man the major approach is by studying the correlation between relatives. One has a set of expectations: for instance, between parent and child one expects a correlation or regression (which, as indicated earlier, is essentially the same thing) to be a half—under simplifying assumptions of no environmental variance, no dominance, and no assortative mating. If there is dominance, or if there is environmental variance, the correlation coefficient will be lower than a half by amounts that can be estimated, while assortative mating will increase the correlation. Also, socioeconomic factors, if they affect the trait in some monotonic way, can have an effect in the same direction.

One very important correlation is that between monozygous twins. They have a theoretical correlation of 1 if there is no environmental effect, since monozygous twins are identical for genotype. There have been many studies comparing monozygous and dizygous twins, where the expected correlations are 1 and ½ respectively under simple conditions. Such comparisons have an important drawback, however, because twins are far from being a random sample of humanity. It is interesting that twins as a whole, whether monozygous or dizygous, have a lower IQ than the rest of the population. There has been an effort to determine whether the cause is prenatal or postnatal, and the evidence is that it is mostly postnatal: it appears that the fact of having to share something in the environment—and therefore having a poorer environment—affects negatively the development of a normal behavioral trait.

One can analyze the similarity of twins in a number of ways. One can measure the correlation or the difference between the two members of each monozygous (MZ) pair and compare it with the dizygous (DZ) twins. Data on stature show that the mean difference of monozygous twins is much lower than that of dizygous twins and that of pairs of sibs. Moreover, the study of twins that have been separated at birth offers a way of partially eliminating the environmental effects of being a twin: data relating to mean differences and correlations between monozygous twins brought up together, monozygous twins brought up apart, and dizygous twins all agree that height of monozygous twins, both brought up together and separated, tends to differ far less than that of dizygous twins. The effect of separation is greater for weight (which one would expect to depend more on environmental conditions) as well as for IQ.

Commenting that the twins who were measured were not in any sense a random sample of the population, Professor Griliches called attention to the problem that these differences will produce an implied estimate of the variance of the residual (which will be referred to as the variance of environment), from which will be developed a metric called the heritability coefficient. The amount of environmental variation depends on the population from which the sample is taken and these groups are not random samples from whatever is the relevant population.

Emphasizing that heritability—a measure that evaluates the effects of inheritance versus those of the environment—is valid only for a specific trait, in a specific population, with a specific range of environmental conditions, Dr. Cavalli-Sforza stated that there are two basic ways of estimating this factor. *Narrow heritability*, the heritability used by animal breeders, is computed on the basis of response to selection. From a population with an initial distribution that is preferably normal one selects a certain number of individuals, usually from the top or bottom, and computes the mean of the selected animals and the mean of the general population. If the environmental variation is nil, and the genes all interact additively (no dominance = allele-allele interaction; no epistasis = nonadditive interaction between different loci), the mean of the progeny of the selected animals will be the same as that of the selected parents; the heritability will be 1. In the opposite case, if there is no genetic component or if it is complex, the mean of the progeny of the selected parents may be the same as that of the entire original population; that is, selection has been ineffective, and the narrow heritability is 0. Narrow heritability depends on the additivity of gene effects, with the heterozygote exactly intermediate (i.e., no dominance). If there is dominance the heterozygote is not intermediate but has shifted towards one of the two parental types. Dominance decreases narrow heritability; epistasis may do the same.

Professor Layzer commented that narrow heritability is strictly an empirical quantity, the ratio between the gain and the selection differential in a selection experiment, and the selection experiment could yield different results when performed with different zones of the population distribution. Dr. Cavalli-Sforza replied that in theory the ratio is constant under certain conditions, one of which is normality of distribution. Dr. Davis commented that in practice one is not dealing with strict normality and so different cuts should indeed yield slightly different figures for narrow heritability. Professor Layzer emphasized the importance of keeping the empirical and the theoretical aspects of the discussion clear from the outset.

Dr. Cavalli-Sforza went on to describe *broad heritability* (which is equal to or greater than narrow heritability), a measure which is computed as the observed correlation between relatives divided by the correlation expected from their degree of kinship. He pointed out, however, that it is very difficult to estimate the expected correlation between, say, a parent and child because it depends on assortative mating, and this is not always estimated accurately and does not produce its effect all in one generation but may take time to reach an equilibrium value. Moreover, estimates of assortative mating are based on phenotypes, and this adds uncertainty. Epistasis and genotype-environment interaction are other complicating factors.

However, in monozygous twins one knows that the expected correlation is 1; therefore, the variance between monozygous twins is entirely environmental and their correlation measures broad heritability. But since monozygous twins are not a random sample of the environmental differences that exist between pairs of individuals, one tries to get a better measure by dealing with monozygous twins that have been separated, with the hope that they represent a random sample of these differences. In theory, the total variance between genotypes may be broken down into several components; but many of these are too difficult to estimate and are therefore ignored. Broad heritability tries to measure the total variance between genotypes divided by the total variance.

In connection with the question of the nonrandomness of the environmental differences of monozygous twins, Professor Feldstein suggested that if similar heritability values were derived from other relationships—parent-child, sib-sib, etc.—one has supporting evidence for the accuracy of the measures derived from monozygous twins. Professor Trivers maintained that even if various heritability estimates within certain kinship groups are the same, this says nothing about the measure of heritability of the population at large, nor about variation between various groups in the population. Professor Herrnstein pointed out that while it is theoretically possible to conclude that similar heritability estimates within different kinship groups are accidental, and that different mechanisms are involved for different groups, it is a rather implausible conclusion. For example, when monozygous twins show about the same values for heritability as other relatives, and show the same overall phenotypic variation as the population as a whole, it becomes hard to argue that these heritabilities are only accidentally similar.

Professor Griliches noted that since heritability estimates relate to a particular population with a particular underlying range of genetic variation and overlying range of environmental variation, and since both these ranges can differ for different populations, one cannot expect heritability estimates to be in any sense constant or extrapolatable to other contexts. Dr. Davis pointed out that no one asserts that heritability has a constant value. Every time either the population or the environment changes one can expect a somewhat different value. However, the magnitude of its variation is not readily predictable: one should distinguish between the theoretical possibility of an infinite range of environments and the actual, empirical limitations on that range. If one finds that some interesting property shows a fairly constant heritability under many different conditions and with a variety of human populations, one should not ignore that fact. Professor Mayr commented that since a calculated heritability is true only for the particular case in which it is determined it is necessary to observe the variation of heritability among different societies or groups within a society; if the variation stays within reasonable limits one could conclude that the heritability value for the particular trait is quite meaningful.

Professor Thomson questioned whether the environmental component is an independent variable—whether it can be assessed independently in assessing the variance in a population. Dr. Davis commented that the environment is defined as the sum total of all the factors that create differences among identical genotypes. In that case, Professor Thomson noted, it is impossible to assess the environmental component independently of assumptions about the heritability. Also, Professor Layzer suggested that such a definition does not account for interaction. Dr. Cavalli-Sforza noted that genotype-environment interaction is an important quantity but it is very difficult to estimate. One would like to be able to classify genotypes in some way, indicating whether specific genotypes will react well or poorly to particular environments, so that the optimal environment could be chosen for every individual. This is the ideal but we are far from attaining it at the present time.

## CONTINUOUS VARIATION IN A BIOCHEMICAL TRAIT

Observing that it may be useful for the study group to think about polygenic inheritance in terms of a biochemical model, Dr. Brewer discussed some biochemical reactions that exhibit quantitative variations between individuals. ATP, a high energy phosphate compound that the red blood cells use for energy, and DPG (diphosphoglycerate), which interacts with the hemoglobin molecule and promotes its release of oxygen in the tissues, are not gene products but rather small molecules that result from the action of many gene products.

They vary quantitatively in level in the cells. This is a typical polygenic situation, where many genes are contributing to some variable which is also influenced by many environmental factors. The human population varies a great deal in the amount of ATP and DPG present in the red blood cells. The mean ATP level of the black population is significantly lower than that of the white population, perhaps as a result of selection pressure because in a malarial environment red blood cells with lower ATP levels provide some advantage. With DPG levels there is the same kind of variability.

In rats one finds three enzymes that affect DPG levels, and one can identify the factors that influence these enzymes. One can then manipulate the level of DPG to make the blood cells more efficient in oxygen transport. The question is whether one can work out the biochemistry of an even more complex quantitative variable. Dr. Brewer felt that it was quite likely that it will be possible to come to grips with some of these problems. He also pointed out that plant and animal breeders are primarily concerned with varying the genes, but in dealing with human populations one is concerned with controlling the environment. Hence identifying the specific environmental factors and their contribution to a quantitative variable may be more important than identifying the genetic factors.

Dr. Brewer observed that the behavioral problem seems to be exceedingly complex because so little is known about the neurochemistry that is involved. However, one can imagine, for instance, that a behavioral characteristic such as anger could involve several neurochemical discharges. Their understanding may make it possible to manipulate the system, so that if the environment is not optimal because of a chemical abnormality this can be corrected. Professor Heymann expressed some hesitation about the suggestion that a state such as anger could be explained by a combination of environmental and genetic interactions. He declared that there are many different dimensions of explanation of behavioral characters like anger—physiological, genetic, psychological, sociological—and it is necessary to analyze the relationship among these various dimensions of explanation. It is frequently objected that this kind of approach is reductionistic, but Dr. Brewer pointed out that, along with analyzing the various parts of the system, one can also carry the analysis to a systems level to understand the interactions of these parts. By a combination of these two approaches one may ultimately come to understand mechanisms that at the present time appear very complex.

## GENE-ENVIRONMENT INTERACTIONS

Dr. Brewer then turned to a discussion of genes in man that are deleterious because of their interaction with novel features of the environment: some have significant overtones in connection with mental disease and behavioral problems. Some are the subject of pharmacogenetics, a field in which a pharmacological agent uncovers the underlying genetic variability. A typical example is a deficiency of an enzyme called G6PD (glucose-6-phosphate dehydrogenase) in the red blood cells. This deficiency is advantageous in a malarial environment, but in its presence certain medicines cause destruction of the red blood cells.

Other kinds of environmental agents may cause serious problems when interacting with genetic factors. For example, there is an inherited deficiency in a protein in the blood called antitrypsin; people with a homozygous deficiency of this enzyme develop the lung disease emphysema. Preliminary evidence suggests that the heterozygous carriers of the deficient gene who smoke cigarettes are considerably more likely to develop emphysema than others. In addition, the frequency of this disease is about two and one half times as great in urban areas as it is in rural areas. It is thought that the particulate matter in cigarette smoke and in the polluted urban air leads to the problem. Since the most frequent causes of death—heart disease, stroke, cancer, and lung disease—show very strong urbanized trends and correlate with the amount of air pollution as well as with other environmental factors, it appears that the environment has a considerable effect on morbidity and mortality. And since all these diseases show some hereditary predisposition, it seems quite likely that the interaction of the genetic and environmental factors is causing these problems.

Changes in environment can also have significant behavioral effects. Studies in animals have shown that overcrowding, for instance, has drastic effects on behavior, and some of the effects—deserting the young, cannibalism, and so forth—are not unlike some of the behavior one sees in very crowded urban environments. This is an oversimplification, but it supports the conclusion that a rapidly changing environment has an important effect on behavior.

Dr. Davis commented on the application of these examples to the problem of dissecting environmental and genetic factors in behavior. If one had proposed 25 years ago that populations who have lived for many generations in malarial regions are likely to be more resistant to malaria than others, it would have been only a reasonable speculation. Now specific hemoglobins and enzymes can be measured independently, and one finds that they are much more frequent in such populations and are correlated with resistance to malaria. However, if today one were to suggest that the higher morbidity and mortality from tuberculosis among American Indians and Negroes than among Europeans is related to the longer history of exposure of the latter to crowded urban environments where a large proportion of the population used to die from the disease, this would again be a speculation. And since invariably there are environmental correlations—the groups with more TB have poorer environments—one can say there is no *proof* that any genetic factor contributes to the higher mortality from tuberculosis among Indians. However, if one could independently measure some gene product that contributed to resistance, there is little doubt that one would find the same kind of correlation that has been found for malaria, for the evolution of resistance to tuberculosis or to malaria should follow the same laws. In the same way with intelligence and other behavioral traits, if one could independently measure some relevant neurophysiological factors the picture would be much clearer.

Dr. Cavalli-Sforza called attention to some data relating to twin concordances for various diseases, pointing out that if there is a genetic factor there should be a higher concordance for the monozygous twins. For tuberculosis there is a 37.2 percent concordance for 135 pairs of monozygous pairs and 15.3 percent concordance for 513 dizygous pairs. This difference is significant and can be interpreted as evidence that there are genetic factors for predisposition to tuberculosis. Several other diseases have been examined in the same way; the one that shows the highest heritability is manic-depressive psychosis.

## EUGENICS

In presenting some historical background for the discussion of eugenics, Professor Mayr observed that the eugenics movement, which began with the work of Galton, was influenced to some extent by some of the concepts of social Darwinism. In its economic aspects this movement was an aberration of evolutionary thought, ascribing to Darwin a way of thinking that he never had. The basic idea was that since man has reached his present state through natural selection an effort should be made to maintain or even increase that level of selection. In Germany from the very beginning the eugenic ideas were expressed in terms of racial differences, which inevitably led to the excesses of the Nazis.

In this country, too, race played a large role in the thinking of many of the early eugenists. They were thinking typologically—this type of person was good, that was bad—and the eugenics movement soon became discredited. At the same time, contemporary movements in other areas made the contrast even stronger: the eugenics movement coincided with the environmentalist emphasis in American anthropology and psychology, where genetic factors were totally ignored; naturally, the eugenics movement, which concentrated on genetic factors, was looked at with hostility.

One must not forget, however, that in spite of the excesses of this movement, some very idealistic people were eugenists. H.J. Muller, whom nobody could accuse of being a racist, had a clear feeling that some people were genetically better endowed and would make a more valuable contribution to society. At an earlier American Academy conference he gave an impassioned speech emphasizing that intelligence was only one of the very large number of valuable human characteristics that included compassion, charity, altruism, etc. He maintained that they all must have a genetic component, and a better species of mankind would result from an increase of these characteristics. When one reads the antieugenic literature one has the impression that eugenics is something that is subscribed to only by racists and others with questionable motives; this is very definitely not the case.

#### Positive Eugenics

Positive eugenics and negative eugenics are very different, both in their methodology and in the possibility of having results. Positive eugenics is analogous to the animal breeder's effort to improve his stock by selecting for traits considered desirable. However, there are considerably more difficulties in connection with an effort to improve mankind. First, one must decide what to select for; improvement of intelligence would be important, but there are many other desirable qualities. Furthermore, animal breeders are interested in uniformity, while diversity is unquestionably a fundamental aspect of human society.

Professor Trivers called attention to two points. First, fitness has connotations that are different from reproductive success itself, but ultimately one always has to discern the Darwinian fitness of a genotype after the fact, by measuring its reproductive success. Second, the species-advantage tradition, which was that of social Darwinists, assumes that individuals within the species can determine what is good for the species in some sense. In fact, individuals will tend, consciously or unconsciously, to compile a list of desiderata that has a bias toward their own genotype; that is how one expects individuals to act if one is viewing them as products of evolution. It is important to recognize, in connection with a discussion about the possibility of manufacturing a better society, that individuals will not be concerned about what is good for the species in some sense independent of their own self-interest.

Professor Mayr agreed that the term fitness has changed its definition several times. Darwin had what a geneticist would call a naïve view of fitness, what the man in the street thinks of as fitness, while the population geneticists later defined it as simply the relative contribution to the gene pool of the next generation. Darwin considered sexual selection responsible for those characters that lead not to the improvement of the species but only to the reproductive success of the carrier of the genotype, while the population geneticists defined fitness as anything that leads to reproductive success of the carrier of the genotype.

Dr. Cavalli-Sforza commented that there is a general belief that many of the differences among races may be the result of sexual selection, a preference for certain types. However, this is very difficult to ascertain because measurements of current preferences—for example, the number of people who prefer blue eyes to brown eyes—do not indicate whether the selection has consistently been in the indicated direction. Although it may well have been very important, there is no reliable way of testing the significance of sexual selection in the human species. He also pointed out that unless one establishes a very strong selection, which is inadvisable for many reasons, the outcome of any eugenics movement, whether it is positive or negative, will be delayed in time for many generations. He expressed the view that a concern with the condition of mankind even ten generations from now is unrealistic in the sense that the likelihood that mankind will reach that future state is not very high unless solutions are found for much more urgent problems. In reply, Professor Mayr pointed out that an effort to solve more immediate problems and an attempt to select for desirable traits in future generations could well be carried on concurrently.

Dr. Davis commented that many people interested in eugenics have put great emphasis on intellectual traits. However, since man now has the power to destroy the whole species, it has been suggested that the greatest danger in our heritage is aggressiveness. Hence a eugenics movement might well select for altruism more than anything else. In response, several members of the group called attention to the potential danger of having a population that was mostly nonaggressive, with a very few aggressive people included in it who could easily dominate the rest.

Dr. Davis noted a more practical problem that might result from the continuation of voluntary population limitation as the answer to the population explosion: if highly educated people are more responsive to that kind of social pressure than less educated people, and if there is some degree of correlation in a fluid society between inborn intelligence and education, there might be an appreciable dysgenic effect in a fairly short time. Dr. Cavalli-Sforza commented that it depends on how fast the idea spreads to the other parts of the population; if it takes on the order of one or two generations, this effect would be very small. In connection with the control of aggressiveness or altruism, he indicated that history demonstrates enormous changes in the average behavior of populations that have been too fast to have been due to genetic change. For example, the Scandinavians are now considered among the socially most advanced people, and yet they probably have essentially the same gene pool as the Vikings, who were the worst pirates centuries ago.

### Negative Eugenics

Continuing his presentation, Professor Mayr stated that negative eugenics is the attempt to reduce the frequency of clear-cut genetic diseases. Some diseases control themselves because the carriers of these gene mutations die before reproductive age, but others do not become manifest until after reproductive age, like Huntington's chorea. There are further diseases caused by recessive genes which can be lethal in homozygous condition; in this case there is the question of whether anything might be done to reduce the frequencies of such genes in the population. There is a naïve attitude that if the heterozygous carriers of such genes could be identified (which can be done in some cases), and if carriers would have fewer children than the other members of their economic stratum, the frequency of these genes would be reduced.

The difficulty is that there is an enormous number of such recessive genes; the statement is often made, although it is probably an exaggeration, that everyone is likely to be the carrier of at least one such deleterious gene. Every time heterozygous parents produce a homozygous child two of these deleterious genes are eliminated from the gene pool; but since this gene is also replenished by occasional mutations, soon a balance is established between the mutation rate and the elimination of the occasional homozygote. Depending on the mutation rate of the particular gene the mutant allele may be carried by one out of two hundred people, or it may be far less frequent. For some highly deleterious genes, like those of diabetes or schizophrenia, the frequency is very high—several percent of the population have genetic factors contributing to the condition.

The question arises how can such deleterious components of the genotype be so common in the population. One answer is that heterozygotes have greater fitness, greater viability in one way or another, than either the normal or the mutant homozygote. We have seen that mechanism operating with sickle cell hemoglobin, malaria, and anemia. In many cases the advantage of the heterozygotes is not known with any certainty, but heterozygote advantage is one of the major factors responsible for the high frequency of genetic defects in the human population. This poses particularly difficult questions for the eugenist: it would not be desirable to reduce the gene frequency to eliminate the disease if, for example, it is found that the gene for schizophrenia contributes to genius in the population.

### Genetic Counseling

Negative eugenics leads to the question of genetic counseling. As indicated earlier, the articulated philosophy of genetic counseling is giving the facts but not giving advice. However, it is difficult to avoid value judgments, and some geneticists tend to go beyond providing the factual information. Dr. Davis suggested that one could argue that a trained medical person is perhaps better able to come to a rational answer in these matters than a person who has not previously been in a position to consider the implications of the problem. Professor Griliches observed that what the rational answer is is not always clear in every case.

Dr. Brewer commented that prenatal diagnosis and abortion is not necessarily effective negative eugenics: for example, if a fetus with Tay-Sachs disease is diagnosed and aborted, the mother is likely to continue to reproduce and thereby maintain the gene in heterozygous condition. Without abortion the homozygote would have been born but not have contributed to the gene pool; and the mother would have been discouraged from further reproduction. Hence the gene frequency may actually go up as a result of this intervention. Professor Brooks suggested that if homozygotes can be identified and eliminated in utero the disease could be eliminated, even though heterozygotes continue to be generated. Professor Herrnstein agreed that if homozygotes can be aborted and if heterozygotes are not at any disadvantage, there is no reason to try to prevent the continuation of the gene in heterozygote condition. Also, one could argue that genetic diversity is something to be encouraged.

Dr. Brewer called attention to another level of the argument, the question of whether a pregnant woman should be required to submit to amniocentesis or whether this would be an invasion of her privacy. Professor Mayr declared that although there are many difficulties and conflicts regarding these matters at the present time, opinions on such questions change over time. Two or three decades ago the approach was very different from what it is today, and one suspects that as it becomes easier to identify heterozygotes, some of the thinking in this area may change even more.

Dr. Davis called attention to another problem relating to genetic counseling in the future: questions are likely to be raised about whether one's conception of the quality of life involves not only the quality of the environment but also the ability of a person to fit into an environment. However much diversity is desirable, there is a level of competence below which a person is unable to function well in a technologically advanced society. Just as attitudes in our society relating to contraception and abortion have changed very rapidly, it may be that within a few decades society will be thinking about extending negative eugenics beyond well defined illnesses, to include persons whose very low position on a polygenic curve make them burdens to themselves. To put it bluntly: an IQ of 60 is equally a handicap whether the cause is a monogenic defect or a normal polygenic inheritance. Professor Mayr called attention to one further dimension of the thinking with respect to negative eugenics: the suggestion that among the rights of a person is the right of a child to be normal. A defective child would therefore have the right to sue his parents for not having prevented his birth.

### Genetic Engineering

In conclusion, Dr. Davis presented a short statement about genetic engineering. There has been considerable comment in the popular press regarding the probability that within a few years the increasing knowledge of molecular genetics will make it possible to blueprint people, and that if that can be done some tyrant is bound to misuse that capability, just as atomic energy has been used in dangerous ways. In Dr. Davis's view, the notion of blueprinting people through altering the genes is not a serious problem in the foreseeable future, for technical reasons. Genetic manipulation in the next century may well make it possible to introduce genes that will cure a monogenic defect; but as we have seen, the interesting aspects of behavior involve an unknown but very large number of unspecified genes—and the cure of monogenic diseases would still be a long way from usefully manipulating such highly polygenic traits.

One technical possibility that is not so remote, however, is cloning, in which individuals are derived, without genetic change, from cells of a single parent. Plants can reverse their control of genes very easily in this way, as in cuttings, but animals cannot. However, frogs have been successfully cloned by transferring the nucleus of a somatic cell into an egg cell, which provides an environment in which the regulatory mechanisms somehow are reversed and all the genes can become active again. The progeny are identical genetically with the donor of the nucleus. It is very likely that before long cloning will be possible with mammals—there will be a strong agricultural incentive to do this with livestock, for one would not have to depend upon the variations of breeding but could produce exact copies of prize bulls and cattle. When that happens it will probably be technically just as possible for man.

Dr. Davis felt that human cloning should be discouraged; it would involve a new kind of moral responsibility, and the emotional consequences of creating individuals with such novel personal relationships to the donors is something that cannot be assessed. Unfortunately, unlike the atomic bomb, which required the investment of a few billion dollars, cloning would require only the mother, a developmental biologist, and an obstetrician. Therefore it probably will happen, but hopefully on a scale that will not represent a serious problem to society.

# Seminar III: Behavioral Genetics (February 15-16, 1974)

## MAJOR PRESENTATIONS

Gerald E. McClearn: *An Overview of Behavioral Genetics*

Seymour Kety, Steven Matthysse: *Genetic Aspects of Schizophrenia*

Lloyd Humphreys: *The Nature and Measurement of Intelligence*

John Loehlin: *Heritability*

*Chairman:* Gardner Lindzey

## RECAPITULATION OF SEMINAR II

Dr. Davis emphasized the relevance, for the purposes of the present inquiry, of the very specific model that molecular genetics has provided for understanding how the environment can interact with the genotype. Genes have controlling elements (operators), adjacent to them in the DNA, which are sensitive to the concentration of chemicals that interact with them; these interactions regulate (induce or repress) the activity of the adjacent gene. In bacteria mutations can alter the sensitivity of the regulatory mechanism to inducers or repressors, and can also alter the maximal capacity of the cell for making the product of a given gene. In extreme cases mutations can make the gene inactive under any circumstances; they can also make its activity independent of specific stimuli. Thus for adaptive enzyme formation we can now understand in molecular terms the basis for differences in sensitivity to stimulus intensity, slope of response, and maximal level that can be reached.

Dr. Davis also noted that at the last meeting there was lack of agreement on what is meant by genetic *capacity* or potential. It is essential to have some such term to refer to *genotypic* differences of intelligence, as distinct from the observed *phenotypic* differences in actual learned material and in developed *abilities.* The model of the immunological system might help to illustrate the distinction. In this system the individual possesses the genetic capacity, residing in many different genes, to make many different highly specific molecules, called antibodies; and the activity of each of these genes is readily quantitated by determining the level of the corresponding antibodies in the blood. This activity depends very much on stimulation by structurally complementary molecules, called antigens, introduced from the environment. Moreover, once this response has been stimulated it persists for some time, as a form of memory. Finally, after initial stimulation a heightened responsiveness to later stimulation with the same antigen may persist for life, as a form of learned ability. Thus an individual faced with the challenge of an infectious disease may solve the problem either by having learned the answer (possessing the required antibody), or by having been trained to construct it (accelerated response), or by relying on an untrained native capacity to respond.

Immunological responses vary with age, nutrition, etc. When these are kept constant inbred strains of mice are found to differ in their capacity to respond to various specific immunogenic stimuli. These differences in capacity include several different quantitative aspects of the response: *sensitivity* to the stimulus (the least amount that will evoke a response); the *rate* at which the response will increase with increasing stimuli; and the *maximal level* that can be approached. These are all aspects of variation in genotypic capacity to make a given kind of antibody.

This model, extrapolated to the nervous system, suggests that the response of an individual to a given set of stimuli, while markedly dependent on his cumulative previous stimuli, also depends on his characteristic set of genotypic behavioral capacities. The methods available unfortunately cannot fractionate the genotypic component of intelligence into sensitivity, rate, and maximal level of response; hence the concept of genotypic capacity refers to a composite of all three (i.e., to the whole curve of response). Moreover, in using this rather undifferentiated concept we are probably dealing mostly with differences in slope of response, and these do not seem as discouraging socially as differences in a hypothetical upper limit. Nevertheless, Dr. Davis suggested that the latter concept is a legitimate part of the picture, even if it cannot be precisely measured: any student preparing for an examination, and anyone trying to perfect a skill, knows the feeling of asymptotically approaching saturation.

Dr. Eisenberg commented on one of the problems of extrapolation: in the immunological model sensitivity, rate, and maximum can be determined by varying quantitatively the exposure of the organism; and one can also compare effects when exposure occurs at birth, at maturity, or whatever. For the nervous system, however, using that model with only one point available would give different information, depending on whether that point was below a threshold, at a maximum, or on an ascending curve. The definition is easy to apply when all three parameters can be determined but it is very difficult when only one measurement is available and the position on the anticipated curve is not known. Dr. Eisenberg also observed that under certain circumstances exposure to an antigen, which would cause an antibody response if presented at a later period, not only causes no response but inhibits forever the capacity to have an antibody response (immune tolerance).

Christopher Jencks noted that the nonspecialist thinks of capacity as a maximum; it should be emphasized that whether there is a maximum in a particular situation is at least questionable. Dr. Davis agreed. The important point is that genetics is founded on the distinction between genotype and phenotype; and in the area of behavior we must make this distinction, whatever the labels we choose, if we wish to apply genetic principles.

## OVERVIEW OF BEHAVIORAL GENETICS

Using concrete behavioral examples, Professor McClearn illustrated some of the basic principles of genetics enunciated at the previous meeting. He prefaced his presentation by observing that many people who are not in the field of behavioral genetics come upon the issue of the heritability of intelligence out of context and may judge that this topic is of overriding interest. In fact, while it is of major concern, those working on the problem are a minority among behavioral geneticists. In the broader context of the field exploration of the whole range of behavioral phenotypes in animals, as well as of other behavioral traits in man, might provide a perspective from which one might better analyze the plausibilities of arguments on intelligence.

In behavioral genetics the general issue of the relevance of animal models arises, as in many other kinds of research, but is perhaps even more acute. There are two polar views: one says that human behavior is unique, so that there is nothing to be learned from the application of animal models; the other is willing to apply immediately and indiscriminately the results of research on animals to conditions of man. Professor McClearn proposed that the proper and defensible position is an intermediate one: the results from animal research may be regarded as being suggestive of the way related factors may operate in man but require confirmation. Since one can conduct experiments with animals that one cannot do ethically with man, such as arrange mating schedules, one hopes that the resulting pool of knowledge will generate hypotheses that are testable on man. All the techniques that are available for human genetic research can be used in animals—twin methods (in those species that twin), adoption studies or models of them, parent-offspring correlations—but the most extensively used approaches have been (1) the utilization of inbred strains and (2) selective breeding.

*Inbred strains* were generated for purposes other than behavioral genetic research, most specifically for cancer research; fortuitously, they differ in interesting behavioral ways. The oldest one, the DBA strain, can be traced back to 1909. Since that time a pedigree control has been exercised in every succeeding generation, thus providing information over a range of generations which would go back, if extrapolated backwards in human terms, to the time of the pharaohs. Inbreeding generally is defined as the mating of individuals with a closer relation on the average than would be provided by a random sampling from the population. In generating mouse strains for research a very intensive kind of inbreeding is employed: brother-sister mating. The consequence of repeated successive generations of sib mating of this kind is to provide a strain of animals that is genetically almost homogeneous.

While certain biological factors probably prevent the attainment of absolute homozygosity, for practical purposes every individual in an inbred strain may be regarded as being genetically like every other, with the exception of the male-female difference in the possession of an X chromosome. Even if there were several heterozygous loci in a typical strain, the existence of segregating loci is of moment only if they happen to be pertinent to the trait being investigated. Because of probabilistic considerations one can assume that no two inbred strains will ever converge upon the same genotype. However, in most cases, except for some rather obvious characters (or for those that have been intensively researched such as immunological systems), little is known about the specific ways that their genotypes differ.

To indicate the magnitude of strain differences, Professor McClearn first called attention to the differences between the means of motor activity scores of two strains, C57 black and A; in fact, the two distributions do not even overlap. Since the two strains were reared and tested under as nearly identical conditions as one can accomplish in a laboratory, the differences in their phenotypic distributions presumably reflect difference in genotype. Differences are also found in the variance of different strains: some have the same mean as the C57 blacks but a much enhanced variability. Evidently, genotypic differences are important in determining not only mean expression of the trait but also differences in susceptibility to the effects of the environmental nuances that cannot be controlled.

As further illustrations of substantial behavioral differences among strains Professor McClearn presented comparative data on maze negotiation, aggression, and alcohol preference. In the last, where the animals were offered a choice between straight water and a solution of 10 percent ethanol in water, abstinence of a very pure kind was found to be characteristic of the DBA animals tested; C3Hs and R3s showed a moderate amount of consumption of alcohol; and the C57 black animals drank about two-thirds of their total daily liquid ration from the alcohol solution bottle. Adult BALBs are alcohol avoiders but at an early age there is a moderate preference for it: a cross-sectional developmental study showed that most of the data points below ten weeks of age were high and those thereafter were low. This raised the question of how BALBs would behave as adults if they had showed a moderate alcohol intake when offered a choice while they were young.

An experiment was conducted in which animals were reared from the time of weaning until the end of their eleventh week of life under one of three conditions: water only, alcohol solution only, or a choice between the two. On the first day of the twelfth week of life they were given a choice and their preferences were observed over a subsequent four-week period. The group that had not been exposed to alcohol behaved just as the adult BALBs previously tested. The animals that were forced to drink alcohol solutions with no other option available showed a substantial increase in alcohol preference: 30 percent of their daily fluid intake was taken from the alcohol bottle. The choice group showed a surprising result: their consumption records indicated that they would drink alcohol for a day or two, then stop, and then drink it again. Moreover, individual animals differed a great deal one from the other. Gradually they worked their consumption up to a 75 percent preference ratio and held it for four weeks. The same sort of developmental study of other strains showed no difference; only the BALBs show this particular kind of developmental change. This suggests a genotype-specific susceptibility to an environmental event.

Genotype-specific environmental susceptibilities were further demonstrated in animals forced to inhale alcohol vapor. In the C57 black and in the DBA groups there is a substantial reduction in activity, in the R3s and the BALBs no significant difference, in the As a hint of an increase, and in the C3Hs a significant increase. In this genotype-environment interaction, which has been replicated several times, different genotypes influenced not only the magnitude of the response to an environmental agent but also the direction—in some cases it enhanced and in others it decreased activity.

Another genotype-environment interaction experiment concerned an effort to determine the effect of gentling—systematic handling of the animals—on the learning process. Two strains, C57 and C3H, were tested on a learning problem; while the C57s that were gentled performed much better than the C57s that were not, there was no difference between the C3Hs. Aggression experiments demonstrate the same phenomenon: when tested under high illumination C57s won 22 out of 25 fights against BALB mice; under low illumination the BALB's won 13 out of 19 fights—the rank ordering of the strains was reversed.

Professor McClearn described an adoption experiment with the AJ strain, who have a very low aggression score, and the CFWs, who have a high aggression score. Animals were appropriately cross-fostered, with the young of one reared by a mother of the other strain, and there was a control, in-fostered group, in which A animals were taken from A mothers and given to other A mothers. The in-fostering had very little influence. The cross-fostering showed an important influence of CFW maternal rearing on the aggression of As, in the direction typical for the mother's strain; but the reverse fostering of the CFW strain to the A mother had no effect at all.

In an experiment dealing with the effects of prenatal environmental influences on the behavior of the offspring, pregnant females of the BALB and the C57 strains were subjected to handling and experiential variation and their young were tested for activity when they became adults. In the BALBs the offspring of treated animals had higher activity scores than controls, but in the C57s the directional effect was the opposite—the offspring of treated animals had lower activity scores than the offspring of the controls. Inbred strain differences have also been identified in courting, social dominance, and audiogenic seizures.

Turning to a discussion of research employing heterogeneous populations, Professor McClearn commented that research using parent-offspring regression analyses with such populations would be particularly appropriate, but very little of it has yet been done. Heterogeneous material is mainly used in *selective breeding experiments*, in which only animals above a certain threshold for the trait under consideration are allowed to reproduce. If there is no heritability for this trait, the location of an individual in either tail or in the middle of the distribution involves environmental events that are not transmissible, and the mean of the offspring will be the same as the mean of the parent group. If, however, there is some heritability, the offspring group will be displaced in the direction of the selection. In laboratories it is the custom to select not just in one direction but in opposite directions, because one can thereby learn a great deal more about the genetic system.

Robert Tryon conducted the classic selective breeding for maze brightness and maze dullness in rats. He devised an extremely reliable maze and bred "bright" and "dull" lines from his foundation stock; in successive generations the distributions of the two populations moved apart as selection was sequentially applied, so that by F20 there was very little overlap. Professor McClearn commented that the labels "bright" and "dull" imply a generality that Tryon himself never meant; there are in fact other kinds of testing situations in which dulls perform as well as brights, and even some cases where brights do less well than dulls. Descriptions of this kind should not be taken beyond their context. Professor Trivers suggested that the terms "more exploratory" and "less exploratory" would be equally descriptive of the behavior.

Emphasizing that the bright-dull labelling is really a besetting trap in the study of animal psychology, Dr. Eisenberg noted that what would be gentling to most humans turns out in many situations to be stress for the mouse, since the effects are duplicated by electric shock. The issue is what is the salient variable: what is the animal attending to, what is the learning process depending on, and is it a generalizable ability or is it highly specific? Professor McClearn agreed: the phenotypes are performance scores, not broken down into components; it is not known to what extent differences in scores relate to differences in some abstract ability to learn, in motivation, in sensory perceptual processing, etc. However, while all of these caveats are important, it is still true that in a large variety of situations the Tryon brights do perform better than the Tryon dulls.

Other types of behavior studied have been autonomic nervous system reactivity, aggressiveness, and sleep time, and selective breeding has demonstrated retrospectively a nonzero heritability in the base population. If that were the only purpose it served it would be most inefficient and uneconomical. However, once animals of this kind are available it is then possible to use them in studying the mechanisms through which genes are exerting their influence on behavior. It is taken as an item of faith that genes operate by conferring specificity on enzymes and that these enzymes are involved in the development and formation of the organ systems of the organism and supervise their functioning, and that therefore the only way that genes can influence behavior is through the nervous system, the endocrine system, the muscular system, the liver enzyme systems, and so on.

Selective breeding for the sleep time phenotype results in strains differing widely in sleep time after administration of a dose of ethanol. Differences in uptake, distribution, and rate of metabolism of alcohol could be eliminated. That left the possibility of a difference either in a central nervous system sensitivity or in the feedback from some peripheral effect of the drug. The next question would be whether this is a generalized sensitivity difference in respect to all hypnotic drugs. The answer is no: the long-sleep animals differ from the short-sleeps for ethanol but not for other standard kinds of hypnotic agents. They do differ, however, for methanol and butanol, so it appears that the animals differ in respect to some sort of sensitivity to the family of alcohols.

Professor McClearn then turned to the evidence that relates *single genes* to behavior. There are generally two approaches. One is to apply segregation analysis to some behavioral phenotype and look at the distribution and mean differences between affecteds and non-affecteds in the F1, F2, and so on; that has been done successfully for a number of traits. The other is to take an already identified single locus for a physical trait and to see if animals differing at that locus differ in terms of any behavioral traits.

Some of the classical examples of the first approach are neurological mutants. A number of single locus conditions have been identified that influence the neurology of the mouse. The first of these to be studied were various behavioral anomalies, such as waltzer (a condition that involves running in small circles) or head shaking. Single genes have also been found to affect mating speed in *Drosophila*, and some think that audiogenic seizures (epilepsy-like fits in response to a noise) may be accounted for by a single gene but that is not entirely clear. PTC tasting in the mouse has also been shown to behave like a single gene trait.

The other approach, beginning with a condition known to result from a single locus and looking for behavioral differences between animals that show the phenotype and those that do not, has also been quite successful. The gene for albinism has been shown by various devices to influence behavior: albino animals are likely to be less active than normals, their learning performance in certain situations is inferior, they consume less alcohol in an alcohol choice situation, etc. This evidence gives reason to hope that even for a polygenic behavioral trait single loci can be identified that are contributing substantially to the variance.

In a brief introduction to human behavioral genetic problems, Professor McClearn observed that there is an important lesson to be learned from the prototypic single locus condition in man, phenylketonuria. Not too many years ago the most extraordinarily heterogeneous collection of characters were grouped under one label, feeblemindedness. Differential diagnoses within the group were not possible until it was discovered that PKU had an identifiable biochemical lesion, so that to the phenotype of mental retardation could be added the phenotype of the presence of phenylpyruvic acid in the urine. It was then possible to pin down the single locus recessive nature of the condition, and research into its biochemistry made it possible to devise a rational therapy, a diet with a very low level of phenylalanine. Galactosemia is another example: if recognized early it is easily corrected by the use of nongalactose-containing substitutes.

Professor McClearn also discussed conditions caused by chromosome anomalies. Down's syndrome, which was discussed in detail at an earlier meeting, results from the presence of one extra chromosome 21. In Klinefelter's syndrome the individual has two X's and one Y chromosome; this results in anomalies of sexual development and some mental retardation, although the situation is complicated because a number of Klinefelters have been found to be perfectly normal intellectually. An interesting condition from the behavioral genetics point of view is Turner's syndrome, which causes anomalies of sexual development in females. In earlier reports these individuals were thought to be moderately retarded, but subsequent work has suggested that in fact their verbal IQ scores are as good as or better than a random draw from the population. They do appear to have a specific disability with respect to dealing with relationships with objects in space. This is a particularly interesting situation in that it indicates a fractioning of the total intellectual domain; it appears that there is not a generalized reduction in intellectual prowess but a very narrow and specific kind of deficit.

In response to an inquiry about research on other animal species, Professor McClearn commented that *Drosophila* have been studied a great deal; selective breeding has been successful for geotaxis and phototaxis, mating speed, and activity level. Steven Matthysse added that Benzer has single gene behavioral mutants in *Drosophila* that have very interesting behavior; these mutants bear some resemblance to human diseases—for example, there is a hyperkinetic mutant, a choreiform mutant, even a mutant that hallucinates. Professor McClearn also called attention to an interesting behavior in strains of bees resistant to an infectious disease called American foulbrood: in the resistant colonies the bees pulled the infected larvae out of their cell and discarded them from the nest, so that a source of further contamination was removed.

One of the big gaps in the animal context of behavioral genetics is in the primate and prosimian area: primate work is expensive, and until the advent of the regional primate centers there was little opportunity for breeding in captivity. Professor McClearn expressed the hope that more attention will be devoted to this area in the future. Professor Mayr pointed out that the length of generations is another inhibiting factor. Professor McClearn agreed but indicated that with a reasonably large population one can conduct parent-offspring investigations; if the regional primate centers were to conduct their breeding in such a fashion, useful information could be obtained.

Commenting that Professor McClearn's presentation dealt in great detail with the use of inbred strains and with selective breeding from heterogeneous populations, Professor Harrison asked why he omitted discussing the classic biometric approach of partitioning variation according to the nature of hybrids. Professor McClearn replied that it has not been very satisfactorily used. A number of studies that used either the classic Mather approach or allele analysis have resulted in much higher estimates of heritability than are obtained either from parent-offspring regression analysis or from response to selection, which is the realized heritability. The parent-offspring approach, using regression and correlational analyses, seems to estimate the realized heritability more accurately than do these other techniques, although this is a matter for some partisan debate.

## GENETIC ASPECTS OF SCHIZOPHRENIA

In discussing the problem of identifying the causes and the mode of transmission of schizophrenia, Dr. Kety pointed out that this problem, though perhaps not as controversial as intelligence, has generated considerable conflict, with significant social consequences. One can define some extreme polarizations. One group maintains that schizophrenia is entirely genetic in origin. This position, taken by many neurologists and biochemists, leads to impatience with any kind of psychotherapeutic approach. The opposite extreme position, espoused by many psychiatrists, social workers, and psychologists, considers schizophrenia almost entirely the result of how parents rear their children. A third point of view, increasingly voiced of late, says that schizophrenia does not exist: it is simply a label applied by the psychiatric establishment to behavior which is unorthodox or of which it disapproves, but which is actually the result of societal inequities. The social consequences of this third view have been disastrous in terms of loss of interest in mental illness, condemnation of methods of treatment, and a tendency to discredit research in psychobiology. These extreme positions are possible because the evidence has been indecisive, and because the problem, like that of intelligence, intersects with social and political convictions.

The traditional evidence for genetic factors in schizophrenia has been essentially that of familial tendency. It has long been known that schizophrenia runs in families, and on the basis of epidemiological surveys one can give approximate figures for the risk. In the general population, in most of the countries studied, it is a little less than 1 percent; in the first order relatives of schizophrenia—parents, children, and siblings—it is about 10 percent, with a range from about 5 to 15 percent; and for monozygotic twins of schizophrenics the risk increases to about 50 percent. However, there are some difficulties in interpreting these data. Assuming that ascertainment is a problem (i.e., that there are more schizophrenics in the population than are ever detected), then a family with two schizophrenics would have twice as much chance of being detected as a family with only one; therefore there would be a selective bias in favor of finding schizophrenics that run in families. Also, the diagnosis is difficult: it really represents an opinion on the part of the psychiatrist, which permits much subjective bias. Finally, since families share not only their genetic endowments but also their environmental interactions, a familial disease is by no means necessarily genetic. In fact, the environmentalists have used exactly the same argument: since the parents of schizophrenics often show a good deal of psychopathology themselves, they have presumably produced schizophrenia in the child by their interaction with him.

One thing that differentiates research in schizophrenia and in intelligence is that schizophrenia traditionally has not been thought of as a continuous variable. Until recently, psychiatrists have felt that it could be described and diagnosed with a fair degree of reliability, even though it is a subjective diagnosis. Schizophrenia has therefore been studied as a disease, rather than as a characteristic that the whole population possesses in a variable degree. But though the population shows a bimodal distribution—with people who are clearly not schizophrenic and others who appear to be schizophrenic—among the latter there is a considerable range in symptomatology. Originally schizophrenia was seen only as a chronic disease with severe manifestations—thought disorder, flatness or inappropriateness of affect, withdrawal, hallucinations—and in Europe this concept is still the predominant basis for the diagnosis. In America, however, that concept has been broadened to include another category, called "latent" or "borderline" schizophrenia, which has the same qualitative characteristics in less severe form, and with greater ability to make some kind of social adaptation. American psychiatry has also characterized an acute schizophrenic reaction, which looks like schizophrenia in temporal cross-section. While approximately one-third of these patients do eventually become more chronically schizophrenic, two-thirds do not.

Because of dissatisfaction with the existing data, Dr. Kety and his colleagues, David Rosenthal and Paul Wender, decided to try to disentangle the genetic and environmental factors in schizophrenia by studies of a total sample of the legally adopted individuals in Denmark between 25 and 50 years of age (i.e., those who had lived through most of the period of risk for schizophrenia). A relatively small number who were adopted by their biological relatives were ruled out. About 14,500 adoptees remained, of whom a geographical subsample of 5500 were first studied. In the national health system in Denmark all admissions to a psychiatric facility are recorded in a central register, and such admissions were found for 500 of the 5,500 adoptees.

The American classification of schizophrenia was used, with chronic schizophrenia, borderline or latent schizophrenia, and acute schizophrenic reaction. The investigators independently read the abstracts of the 500 hospital records. If all agreed independently that the patient was a schizophrenic then he became an index case; if all agreed that he was not, he was discarded; and if one investigator thought that he might be and was able to convince the others, this individual, too, became an index case. Disagreements were largely over the borderline schizophrenics and the acute schizophrenic reactions.

This procedure yielded 33 index cases. An equal number of controls were selected from the rest of the 5,000 adoptees: they had no known history of mental illness and they matched each index case in age, sex, socioeconomic class of the rearing parents, and length of time spent earlier with the biological family. For these 66 individuals the biological and the adoptive relatives were then identified: mothers and fathers, full siblings, and half-siblings. For the resulting 512 relatives psychiatric registers and military, court, and maternity records were searched for information about psychiatric illness, extending the spectrum of schizophrenia-related illnesses to include a category of doubtful schizophrenia and a category of schizoid or inadequate personality. As with the adoptees, the records were diagnosed by forced consensus of the raters. After diagnosis the code was broken and the relatives were allocated as biological or adoptive, and as related to index or to control cases.

The biological relatives of the index cases did not show an elevated frequency of mental illness in general. However, they did show a very significant concentration of disorders of the spectrum of schizophrenia: sixteen of the total biological relatives of index cases plus controls were identified within the schizophrenia spectrum, and thirteen of these were relatives of index cases. Moreover, the adoptive relatives who had reared a schizophrenic showed the same low incidence of schizophrenia as those who had reared a control. When the survey was later extended to include all of Denmark, with a total of 74 adopted future schizophrenics, these findings were confirmed.

The relatives were further studied by interviews, to detect possibly relevant mental illness and personality disorders, and to look for differences in the environments that yielded schizophrenics and those that failed to do so. The interviews included an exhaustive mental status examination. Again, while biological relatives of the index cases did not show an excess of psychiatric disorders other than schizophrenia, 24 percent of them were diagnosed as being in the schizophrenia spectrum, as opposed to 7 to 12 percent of the other three groups of biological and adoptive relatives. For definite schizophrenia a similar concentration existed: 8 percent among the biological relatives of schizophrenic adoptees, and 1 to 3 percent among the others. Incidentally, the interviews revealed three times as many people in the schizophrenia spectrum as had been seen by psychiatrists—an indication that the prevalence of schizophrenia and related disorders is even greater than had been generally realized.

Dr. Kety commented that these data, though quite compelling, still do not unequivocally demonstrate that genetic factors are responsible: the separation from the biological family took place a week to a month, sometimes even a year, after the birth of the child; and in utero factors might have played a role. Talcott Parsons pointed out also that the adopting environments were by no means uniform. Dr. Kety agreed but indicated that serious selective bias in placement was not likely, for where mental illness did occur in the adoptive parents it was found on the average twelve years after the adoption. Furthermore, although the adoptive environments might affect the incidence of schizophrenia in the adopted children, its incidence in their biological relatives could not be affected by these environments. It is also relevant that the incidence of schizophrenia is the same in the biological relatives of adopted and of nonadopted schizophrenics; and the overall prevalence of the disease in the total population of 5,500 adoptees was the same as in a matched group of nonadopted individuals.

To test for a possible influence of early mothering and of in utero environmental factors, as well as of communication of information about the disease in relatives, Dr. Kety presented some data on paternal biological half-siblings. Among the 512 relatives 63 such half-siblings were found for index cases and 64 for controls. Among these 127 individuals there were sixteen instances of schizophrenia, and fourteen of these were in relatives of index cases ($p = 0.001$). Since biological paternal half-siblings do not share the same in utero environment or early mothering experience, and after early adoption neither member of the pair is likely to know about schizophrenia in the other, this finding is indeed compelling evidence for a major contribution of genetic factors in the development of schizophrenia. Moreover, while the concordance of only 50 percent in MZ twins suggests that a person with the susceptible genetic structure will often not develop schizophrenia unless it is triggered by some nongenetic factor, that unknown factor does not seem to be substantially correlated with the in utero or the early mothering environment, nor does it require rearing in a family prone to schizophrenia.

Dr. Matthysse then briefly discussed what is presently known about the mode of transmission of schizophrenia: which chromosome the relevant genes are on, and what proportion of the variance is contributed by the inherited factors. While knowledge in this area is still extremely limited, single genes can sometimes be found that are relevant to a condition even if the mode is not fully known and even if there are complex polygenic contributions. The most encouraging result in this field is the recent evidence for X-linkage in manic-depressive illness. Although the evidence is not conclusive, recent work by Winokur and his colleagues shows a relative deficiency of father-son transmission. Also, there were discovered some families where colorblindness (located on the X chromosome) was segregating with manic-depressive illness, and all the family members with the affective disorder were either colorblind males or carrier females.

Once one finds putative linkage to a marker situated on a known chromosome, then other markers on that chromosome can be examined intensively and the study becomes more focused. Therefore, other investigations have concerned Xga, a blood group marker on the X chromosome, and several pedigrees have been found where Xga is also apparently linked to manic-depressive illness. Unfortunately Xga and color blindness lie on opposite ends of the X chromosome, which raises some doubts because it is unlikely that the disease could be linked to both of them at once since they are not linked to each other.

No one has seriously suggested X-linkage for schizophrenia, and autosomal linkage is much more difficult to identify. Dr. Matthysse analyzed the chances of finding autosomal linkage in schizophrenia. It is generally believed that if the map distance is more than 20 centimorgans (1 centimorgan representing a 1 percent chance of crossing over) linkage is extremely difficult to establish. With that as an upper limit for distance, one can ask what is the probability that out of, say, fourteen markers distributed over the chromosome set one of them lies within 20 centimorgans of a gene that contributes to schizophrenia. That probability can be shown by computer analysis to be 17 percent. At the present time there are 28 markers that are readily available in a good blood typing laboratory. However, of these only about ten are polymorphic enough to be very useful, so the probability of success lies somewhere in the low teens. This is encouraging, although it is by no means certain that linkage will be established.

Establishing linkage has many values. It is very helpful in ascertaining the mode of transmission. If a linkage is close one might be able to use it for prenatal detection. Sometimes close linkage suggests a functional relationship, especially among the immune response genes—that is, diseases closely linked to the HL-A histocompatibility gene are sometimes diseases of the immune system or diseases involving autoimmune processes. The most exciting potential use of linkage, although many years in the future, is that once one knew which chromosome was involved one could study the defective chromosome in culture by fusing human cell lines with either mouse lines or other human lines and choosing media in which that particular human chromosome would be expressed.

As indicated earlier, many different genetic syndromes can produce mental retardation, which behaviorally could easily be taken to be one entity. Schizophrenia might similarly have several genetic types in the population. One advantage of pedigree analysis is that it is much more likely that one genetic type is prevailing within one pedigree rather than in a heterogeneous collection. There is not a great deal of knowledge about genetic types in schizophrenia. However, one interesting study might be followed up: some years ago Dr. Kety showed that about 20 percent of schizophrenics become very much worse when given large oral doses of methionine, while the others do not. It has since been shown that this is a replicable phenomenon—a reactor to methionine continues to be a reactor and a nonreactor seems to remain nonacting. Conceivably this is only a matter of pharmacogenetics, but it might be of interest as an indicator of genetic subtypes.

It would obviously be of great value to have a biochemical marker of any kind for schizophrenia. There is not much that is encouraging, except that platelet monamine oxidase has been shown to be low in schizophrenics. (It has also been found to be depressed in manic-depressive illness, but not as much.) Since platelet MAO is also decreased in discordant monozygotic twins of schizophrenics, it may be a genetic marker.

A second question is what proportion of the variation in schizophrenia is inherited. As was stated by Dr. Kety, about 50 percent of the monozygotic twins of schizophrenics are schizophrenic. These twins normally share their rearing environment as well as their genes, so it is often concluded that the variation must be at least 50 percent environmental and maybe even more. However, in Dr. Matthysse's view, the logic involved in this argument is wrong, and he indicated that the same logical analysis also applies to all questions of concordance versus heritability in quantitative traits like intelligence. The first question a hereditarian might ask about the 50 percent MZ concordance is how normal are the discordant twins. While they may not be schizophrenic, they could have some mental illness. However, studies indicate that many of them are quite normal. Another possibility is that a discordant monozygotic pair indicates a phenocopy, a case of schizophrenia caused by infection or environment—something other than genetics—while a concordant pair would indicate the genetic schizophrenia. That would be very important if true because it would be a way of selecting from the population of schizophrenias those which were environmental phenocopies and those which were genetic. However, that is not the case either.

Dr. Matthysse referred to a study by Margaret Fischer, who looked at the incidence of schizophrenia in the offspring of normal cotwins of schizophrenics. If the schizophrenia in the sick twin was a phenocopy—that is to say, something that resembles the genetic illness but is caused by environment—the cotwin would have no genes for schizophrenia to transmit to his progeny. Nor is there any reason why his environment should be particularly pathogenic, since he is not psychiatrically ill; therefore the incidence of schizophrenia in the offspring of the normal cotwin should be the population incidence, about 1 percent. But it is not (the incidence is 12.3 percent), so the discordant twins apparently do not represent phenocopies. One might say that the environmental influence could be prenatal, and although prenatal causation may seem unlikely, the number of genetic diseases with profound prenatal contributions is so large that it cannot be discounted. In cleft lip and palate, for example, which is detectable at birth and therefore has no postnatal influence, monozygotic concordance is 37.7 percent, very close to that of schizophrenia, and yet dizygotic concordance is only 8 percent, so it is clearly an inherited disease.

Professor Feldstein commented that a large role of environment is not so surprising if the disease only becomes apparent later in life and if the environments move apart substantially during the years before the disease actually manifests itself. Dr. Matthysse agreed, and he noted that most people in the field conclude that there must be a very sizeable postnatal environmental influence because the concordance rate is 50 percent or less and all the other possible explanations do not apply. He emphasized that 50 percent concordance does not mean 50 percent heritability. He described some mathematical models based on monogenic modes of transmission where heritability estimates as low as 8.2 percent (with a recessive model) or as high as 98.5 percent (with a dominant model) can be derived consistent with a concordance rate of 40 percent and a general incidence of 1 percent. Calculations with a polygenic model result in a heritability estimate of 80 percent for that concordance and incidence (C. Smith,

"Heritability of Liability and Concordance in Monozygous Twins," *Ann. Hum. Genet.* 34: 85-91, 1970). Since both single gene and polygenic models are compatible with the available data, it is not yet possible to determine whether the gene-environment interactions in the transmission of schizophrenia are analogous to those in intelligence or whether they present qualitatively different phenomena.

Professor Mayr commented that in a recessively inherited condition there will be a significant difference between the incidence among the proband's sibs and that among his parents; that is, there is a very low frequency among the proband's parents, while among his sibs it is about 25 percent. In schizophrenia the frequency among parents and among sibs of probands is about the same and therefore many authors have utilized a dominant model. He asked whether there was any evidence for considering a recessive model in this case. Dr. Kety observed that while some pedigrees in his adoptive study are most consistent with dominant transmission, they represent a limited subset of the total sample. If in fact there are different kinds of schizophrenia, they may have different modes of transmission.

## HERITABILITY

As introduction to his presentation on heritability, John Loehlin reviewed the distinction between polygenic and monogenic traits. Monogenic traits primarily reflect conditions at a single genetic locus. If one allele is present at that locus on both chromosomes of a pair the organism shows one phenotype; if another allele is present at both loci the organism shows another phenotype; and if different alleles are present on each chromosome the organism may show the first or the second or some intermediate phenotype, depending on whether one or the other or neither allele is dominant. Thus individuals will fall into either two or three distinct phenotypic classes, depending on whether dominance is present or not. Many traits, including a number of behavioral ones, behave in this simple Mendelian fashion. For a polygenic trait, whose development is influenced by a large number of chromosomal loci, and perhaps with numerous alleles in the population, each individual gene behaves in exactly the same simple fashion as in the monogenic case; but their joint effect is to produce a large number of different phenotypic classes representing all the possible combinations of these genes.

Estimating the components of variance is the basic process that underlies heritability analysis. *Variance* is the average squared deviation from the mean of individuals in a population. Professor Loehlin pointed out that it is useful to remember that disproportions between heredity and environment are more marked on a scale of variance than they would be on the scale of original measurements. It is also essential to keep in mind that the variance is a property of a population; it describes the way in which individual scores are either grouped together or spread out in the population in terms of the characteristic in question.

If the variation of a trait measured by the variance is due to the action of several independent influences ($V_A$, $V_B$, $V_C$), the total variance can be regarded as the sum of the variance due to the separate influences ($V_A + V_B + V_C$). This assumes that A, B, and C are uncorrelated in their occurrence and independent in their effects on the composite trait, which in real life usually is not true; sometimes it is not very far from the case and an approximation of this kind will work fairly well, but in other instances it will not.

To take such factors into consideration one must insert additional components into the right-hand side of the equation: if A and B are correlated but independent of C, this can be dealt with by adding a covariance term to account for the additional variance produced by that correlation; or if A and B act jointly to influence the trait an additional term, A $\times$ B, can be added to represent interaction. A, B, and C might represent variance due to the genes, variance due to the environment, and errors of measurement; or alternatively they might represent variance due to social class, variance due to family, and individual differences—or any other way of dividing up influences.

Professor Layzer commented that if $V_A$ is the variance that would be present if the other factors were held constant, the question arises of the values at which they are held constant. If the environment is changed keeping genotype constant, over the range of environments is $V_A$ the same as it would be if some other genotype were held constant? Professor Loehlin replied that if in fact the effect of the genotype differs at different levels of environment, then what is used is essentially an average effect of genotype over all environments or environment over all genotypes in the population. While this is not a strict Fisherian definition it is based on his mathematical approach. In Professor Layzer's view that is a more sophisticated approach than the comparatively naïve method, which is to simply take some fixed mean values of the environment and the genotype of the population and consider that those were the reference values from which $V_A$ and $V_B$ are calculated. That would yield quite a different kind of partitioning of the variance. One has to keep in mind the possibility that $V_A$ and $V_B$ may be negligibly small compared with the $V_{A \times B}$; that would be a sort of "worst case," but it is not clear that that is not the case for all behavioral traits.

In response to a request to define what were referred to as the naïve and the sophisticated views, Professor Layzer observed that if the genotypic variances for Environment A and for Environment B are the same there is no problem. If they differ the sophisticated view (the Fisher definition) would be that the genotypic variance is the appropriate weighted combination of these two variances, depending on the frequency of occurrence of those environments in the population. The naïve view would be to take one of those as a reference and define the genotypic variation as the variation that occurs at that environment.

It might be simpler to take that particular model and define both environmental and genotypic variance with respect to reference genotypes and environments, but in that case genotypic variance is purely genetic and environmental variance is purely environmental, and the disadvantage is that it puts more of the variance into the other terms that are difficult to estimate.

**Components of Phenotypic Variation**

Professor Loehlin continued that for purposes of the present discussion three categories of possible variance components are distinguished: components of the variation that are genetic ($V_G$); components of the variation that are environmental ($V_E$); and components that are not clearly assignable to either of these categories: correlation between genotype and environment, interaction between genotype and environment, measurement error, and others. Within each of these categories various subcomponents can be distinguished. Within *genetic variance* one can distinguish variation due to additive effects of genes (the average effects of relevant alleles across all the genetic contexts in which they occur in the population), the additional variation due to dominance (relationships between alleles at a particular chromosomal locus), as well as that due to epistasis (the interactive effects of genes at different loci).

One can also allow for the effects of assortative mating if this factor is related to the trait in question. Or one can divide the genetic variation in other ways, such as genetic variation within families and genetic variation between families. The relationship between a within- and between-family variance component is a function of these other parameters—dominance, assortative mating, and so on. In a particularly simple case, random mating and entirely additive genetic variance, the within- and between-family components of genetic variation are equal. Positive assortative mating tends to raise the between-family component relative to the within; the presence of dominance or epistasis acts in the opposite direction by lowering the between relative to the within; and if both of these factors are present, as they probably are in a trait like intelligence, the within- and between-family components are likely not to be drastically different.

The *environmental variance* can also be split into between- and within-family components. Members of a particular family have certain environmental influences in common. For example, during any given period they tend to be living in a certain neighborhood, share a certain level of poverty or affluence, have various kinds of common institutional identifications. At the same time members of a family will also differ in environmental influences—individuals will occupy different roles within the family, they will interact with somewhat different sets of peers, they will have different experiences at school, and so forth. It is less clear in the case of environment what the ratio of common to diverse influences might be; it seems likely that the common environmental influences will be greater for siblings than for parents and their offspring but beyond that it is hard to do more than guess at the relative relationships.

Professor Layzer commented at this point that when one speaks of the genetic component of variance he is talking about the genetic component for a given distribution of environmental factors in a population, and the definition of the genetic variance will change as the distribution of environments changes; the same is true for the environmental component of the variance. If Fisher's definitions are adopted genetic variance should have an explicit reference to the environmental distribution built into it, and conversely. Professor Loehlin asked whether Professor Layzer was referring to the fact that one is describing a particular population, with a particular range of genotypes and a particular range of environments. If so, the genetic and environmental variation are implicitly with reference to this population; the ratio would of course be different for another population, with a different range of environments and a different range of genotypes.

Professor Griliches felt that the point was that the range of expressions observed for a particular trait, for which one does not have direct measures of the genes, is conditional on the range of environments—for example, if there were no milk one would not know there was genetic variation for lactase. One is dealing with a set of observations and with an interpretive model that has a certain power of predicting some other associations, but it involves a variety of simplifications because there are no independent measures. If one could define genetic variance in terms of identifiable alleles, or of differences in amino acids, or if the relevant environmental variance could be measured in terms like centigrade temperature, one would have independent measures of variance. Without such independent measures it is difficult to define what each variance is.

Using the example of height as the variable of interest, Professor Feldstein commented that with an exactly identical environment for the whole population the observed variance would be all genetic, but it would differ depending on the environment, because some environments would bring out more of a range of the underlying variance. If these individuals were placed in Environment A they would have a different measured variance than if they were placed in Environment B.

Commenting that the previous remarks seemed to reflect an effort to deal with differences between populations or between subpopulations in a single model, Professor Loehlin indicated that his interest is not in such comparisons but in analyzing the variance components for some population and seeing how these might change if the environment changed or the genotype changed. Professor Layzer emphasized that with reference to the standard definition the absolute value of the genotypic component of variance would change if the distribution of environments in the population changed.

Professor Feldstein pointed out that if the broad population was divided into environmental groups and it was discovered that the variance is the same within each, that would indicate (if the subpopulations were genetically similar) that one need not be concerned with interaction. But if it is discovered that the variance is different within a particular population depending upon the environment, then one would always have to take this into account. Professor Layzer added that in the same way, if it was discovered that environmental variations did not depend upon genotype in their effects—for example, if the variance in improvement were the same for all six-year-olds given X hours of taped instruction in reading—then one need not worry about this sort of effect.

Professor Loehlin defined the third set of factors that influence the variance of a trait, those that are difficult to assign either to the genetic or to the environmental component. One is *gene-environment correlation*, of which there are two types. One type is due to the fact that parents transmit to the offspring both genes and environments, and these may be correlated; intellectual parents may give their offspring genes that facilitate the development of intellectual behavior and also provide them with environments that tend to do likewise. The other kind of correlation is that due to the behavior of the individual himself in selecting environments that tend to affect development of the trait, i.e., bright children spend more time in libraries than less bright children. This will bring about a correlation between the genes and the environment which is not imposed on the individual from the outside but depends on the individual's producing this correlation in his interactions with the environment.

In the case of intelligence positive gene-environment correlations are discussed for the most part. However, negative correlations are possible, and with personality traits they may in fact be important. For example, dominant parents might transmit to their offspring genes favoring dominant behavior but provide them with environments that are conducive to the development of submissive behavior; this would result in less total variance, due to the fact that certain tendencies of the genes are being repressed by certain environmental pressures. In areas of personality traits there may be a good deal of socializing behavior that is designed to restrict the range of expression of certain traits, resulting in less total variance.

*Gene-environment interaction* also contributes to the variance and also cannot simply be assigned to genotype or to environment. A component of this kind is present whenever a particular combination of genes and environments has effects that are not predictable from those genes and those environments considered separately. There may be extremely favorable or extremely unfavorable combinations, where certain genotypes may be ineffective except in particular environments; there may be different optimal levels of environment for different genotypes; there may be thresholds on either the genetic or environmental side below which nothing happens and above which the trait is manifested.

Another source of genotype-environment interaction is scaling—that is, there may be trait effects which are additive on a logarithmic scale but not on a linear scale. This kind of nonadditivity can be included in an interaction component. In this situation the normal reaction of biometricians is to revise the scale to eliminate the interaction. However, if one is interested in describing an existing trait as it is measured, then rather than rescale it to eliminate the interaction, one might want to retain an interaction component. Professor Layzer commented that one is only able to do this under conditions in which a function of N variables can be transformed, by suitable transformations of those variables, into an additive function, and the a priori probability of being able to do that is zero. Professor Loehlin suggested that the practical probability is much greater than that because very often the nonlinearities of scale are simple monotonic ones.

Professor Layzer remarked that in addition to analyzing interaction from the causal point of view, there is a more empirical approach which animal and plant geneticists use: one can consider a given sequence of genotypes and look at the effects of varying environment over the range of interest for each of these genotypes, or one can take a sequence of standardized environments, pick out the relevant environmental factors, and look at the results of varying genotype for fixed environment. The standard test then of the importance of interaction is the extent to which these distributions differ. Professor McClearn pointed out that while interactions of this kind can be demonstrated in inbred animal strains, where one can identify several genotypes, apply several different environments, and obtain interactions, one cannot order the genotypes—it is a strictly nominal scale—nor in most cases does one know enough about the environments to order them.

Professor Harrison commented that one can check the variances of the two parents and of the F1 as a measure of environment-genotype interaction, and one can scale for interlocus interaction by predicted relationships between hybrids. Professor Layzer maintained that one does not actually have to scale things to determine whether an interaction is present or not. For example, the alcohol preference experiments showed that different strains did have quite different variances exposed to the same environmental conditions. In that case, whether one can order the genotypes or not, interaction is very large, and this kind of decomposition would give small factors for $V_A$ and $V_B$ and a large factor for $V_{A \times B}$.

Dr. Davis suggested that the self-selected covariance is in a sense already built into the arbitrary IQ scale. That is, if one determines the number of questions that each person answers correctly and then sets distances on the scale in a manner to provide a normal distribution curve, to the extent that the people in the upper part spend more time in libraries and therefore sharpen the higher intellect they were born with, with the opposite occurring on the lower part, that may simply be an amplification of the genotypic effect that gets built into the scale.

Professor Loehlin did not believe that it is a function of the measurement. He agreed that it is very difficult to distinguish between the self-selective correlation component and the pure genotypic effect—in fact in the last analysis genotype only has its effect by selecting from biochemical environments—but most psychologists at least would like to distinguish between that kind of selection of environments and the kind that involves interacting with other people, shifts within social institutions, and so on. Dr. Davis suggested that the presence of environmentally imposed covariance complicates but surely cannot eliminate any demonstration of genotypic variance, as Professor Layzer has suggested, because if there were no genotypic difference from the mean there would be nothing for the environment to correlate with. Interaction seems to be the most serious problem.

Professor Loehlin observed that in some definitions the genetic factor would relate to what the DNA does and the environmental factor would include everything else; measurement error would then be included in the environmental component. In his view, however, the deficiencies in the "ruler" ought not to be apportioned either to the genotype or to the environment of the measured object but should rather be kept as a separate component of variance, perhaps excluded before further analysis is begun. The more components one understands the better, but not much is gained by arbitrarily assigning them to either the genetic or the environmental side.

Another aspect of the third category of components contributing to variance is that of changes in heritability with age. If one measures siblings at the same age one may very well find them more like than if one compares them at different ages, or at another age. By some methods such differences would be classified as environmental because the genotypes are not different at different ages. However, the developmental trends that differ from the standard curve for development with age may be partially under genetic control. Since the situation is ambiguous, in the absence of further data this component should also either be considered as separate or be eliminated from the analysis.

Professor Jencks commented that since one does not know that the genes that affect performance when tested at one age are the same as the ones that affect it at another age, one does not know that the genotype is constant in any real sense. If one allows for that uncertainty, one would not be able to refer to the heritability of IQ but only to the heritability of IQ at a certain age. Professor Loehlin agreed that one could very reasonably take that position for heritability at age 2 versus heritability at age 4. Dr. Davis suggested that such measurements at age 2 should properly be called something other than intelligence tests because they are tests of motor performance not involving abstraction, and they correlate so poorly with later IQ tests.

Professor Jencks also commented that one would logically expect to find heritability maximal during the period in which people were in school, since their environments are relatively uniform at that period; if one were to measure their IQs at the age of 60, one would think it at least plausible that there was considerably more environmental variance affecting their vocabulary between the ages of 20 and 60 than between the ages of 6 and 18.

**Estimating Components of Variance**

Professor Loehlin then turned to the question of how one estimates such components. With inbred strains of mice it is relatively simple: one has a selection of genotypes and environments; he can correlate them or not as he chooses; and he can measure the genotype, the environment, and the interaction components. The major question is whether the genotypes or the environments—and hence whether the results—generalize to any population of interest.

In the case of humans it is much more difficult. One can subdivide the components of variance infinitely in theory but in practice very few researchers have attempted to do more than estimate a few of them, and those often under somewhat tenuous assumptions. Some components are easier to estimate than others. Repeated measurements of the same trait on one's subjects can provide an estimate of temporal trends as well as an estimate of error of measurement or fluctuations around those trends. The genetic component is usually estimated by finding pairs of individuals who have different degrees of genetic resemblance and what are hoped are comparable degrees of environmental resemblance. The classic design is comparison of identical and fraternal twins, identical twins having much greater genetic resemblance than fraternal twins but degrees of environmental resemblance being more or less comparable. Alternative methods include comparisons of the variances of biological siblings with those of unrelated individuals reared together by adoption, or of siblings adopted into different families with members of different families who are not related to each other.

Professor Layzer noted that when animal geneticists apply such techniques to estimate the components of variance they invariably take pains to randomize environments. In these circumstances the estimates are theoretically well founded even when the interaction component is large—and one expects it to be large for phenotypically plastic traits. This randomization explains why animal geneticists have been moderately successful in getting agreement between heritabilities estimated from selective breeding experiments.

When that condition of random environments does not obtain, as it does not for most human subpopulations, the theory is on a completely different qualitative footing, and so far as one can tell from looking at the mathematics it simply does not apply. In spite of the superficial analogy between the animal case and the human case, unless one can find natural experiments in which the environments of the subjects are uncorrelated with their genotypes to a high degree of approximation, one cannot apply this mathematical machinery to make the connection between measured correlations of behavioral traits and heritability.

Professor Mayr commented that while it is not possible to use the methods employed by animal breeders in estimating heritability, this does not mean that other measures of heritability have no validity. There are two sets of approaches: one can either compare identical genotypes like MZ twins with other less closely related individuals in the same environment, or one can change the environment by adoption experiments with separated MZ twins or with unrelated persons. People are trying in various ways to cope with the problem of estimating the correlation between genotype and environment. To be sure, estimates based on the considerable differences in performance of fraternal twins raised in the same environment as compared with MZ twins have an error—and it may even be of considerable magnitude—but they still provide a measure that is quite meaningful.

Professor Layzer expressed his agreement with the qualitative conclusion that these correlations show that genetic factors play a very important part in development; when genetically closely related individuals are tested in similar environments they turn out to perform similarly, and the closer the genetic relationship the more closely correlated the performance. However, he maintained that such estimates depend on arbitrarily excluding factors that are of the same order of magnitude as those which are included. When one goes from the level of plain language to the quantitative level, the a priori probability that this analysis could yield meaningful results, in a situation where there are large neglected terms of the order of those that are included, is negligible.

Dr. Davis asked Professor Layzer why, if he agreed with the qualitative conclusion that there are large genetic components contributing to people's behavior, when he examined the mathematics he came to the conclusion that the results are worthless. A more reasonable conclusion would seem to be that because of the uncontrollable or undeterminable variables the precision of the answer is lower than one would like but still has some meaning. Professor Layzer stated that the general conclusion is that both genetic and environmental factors are very important and that small environmental differences will give rise to small phenotypic differences when the subjects are closely related genetically. But the issue that arises when one tries to translate these into numerical terms, and into terms of heritability, is whether there is a purely genetic component and a purely environmental component. In that equation one conceivable situation is that in which $V_G$ is equal to zero, $V_E$ is equal to zero, and only interaction and correlation are significant.

Professor Mayr maintained that while it would be ideal to have exact quantitative statements, as long as one knows that there is a large genetic component as well as a large environmental component, the fact that one cannot measure them exactly, while regrettable, is not of major significance. Dr. Kety emphasized that no one is denying the existence of a genetic component and an environmental component; the question Professor Layzer is raising is that to attempt to give a precise figure for heritability on the basis of the evidence that has been suggested at the present time is spurious, because one is excluding variances that are as large as variances that are being included. With regard to comparisons of MZ twins and siblings, it is a valid hypothesis that the environments are much more congruent for MZ twins within a family than they are for siblings or for DZ twins. One has to take that variance into consideration; one cannot just assume that it is small.

Professor Griliches pointed out that in analyses of nonexperimental data assumptions are made that are in some sense not testable directly on that body of data, and that in general can only be interpreted as conventions for the analysis. If one does not know what genes and environment are he may want to assume, in the absence of evidence to the contrary, that the two are uncorrelated and proceed with the calculations on that basis. If at some point an explicit measure of the environment is obtained it can be put into the calculations to check whether the original assumption was correct; and if not, one would then reanalyze the data appropriately. But in the absence of specific alternatives there is no way that one can test and reject a generalized assertion that all of one's calculations are spurious because the assumptions are wrong. In analyzing nonexperimental data one has to assume that the excluded factors are random with respect to the factors that are being analyzed. This need not be true in any one particular case, but to debate the point is useful only in the context of an explicit alternative testable hypothesis. Then it is possible to check whether the results are robust or not.

Professor Layzer observed that the situation is the same in all of the observational sciences. The point is that genetic and environmental factors may each be contributing very importantly to phenotypic differences in a variety of ways; virtually all of the variance may be due to gene-environment interaction or there might be a very small interaction contribution. The only way of settling that issue, which is a strictly quantitative mathematical one, is to look carefully at the mathematics and errors of the analysis. There is no a priori reason why a formula that works in one situation should yield even approximately meaningful results in another.

Professor Feldstein asked what factor is excluded in the hypothetical experiment of looking at the difference between MZ and DZ twins raised in the same family, including families in which the DZ twins are not recognized as such so one cannot hypothesize that they would be treated differently. Professor Layzer suggested that since little is known about what environmental factors affect the kinds of abilities that are tested by IQ tests, there might be any number of environmental factors that would be very difficult to control.

With regard to the comment that errors in heritability measurements are very large because the environmental differences between the nonidentical twins are much greater than between identical twins, Dr. Davis pointed out that the overall range of environments of the total population varies from the worst slum to the most enriched environment, but within any one family the range is very much smaller; hence the difference in the environmental variance of identical and nonidentical twin pairs, while real, is not likely to be a large part of the total $V_E$.

To focus the discussion, Professor Loehlin called attention to some specific figures used in estimating heritability, together with some of the implications of different methodological problems. One such figure is the correlation between monozygotic twins reared apart in separate families, for which the median value given by the Erlenmeyer-Kimling study is .75. However, it was pointed out that there is an important question as to how different the environments are in those studies, especially since in many cases only one of the pair was adopted while the other was retained by his biological family; with any significant amount of selective placement the environments may not be different enough to bring out the ordinary random environmental variance. Also, the survey was based on four studies, and some of the samples are very small. Therefore one cannot assume that there is any generalizability to these figures.

Dr. Davis commented on the argument that since some of the separated monozygotic twins have been raised in the homes of relatives, rather than in random environments, much of the high correlation of .75 would be artificial and due to the reduced range in environment. This argument implies that two different families that have some relationship represent only a small fraction of the whole possible range of environments. On the other hand, if within a family monozygotic and dizygotic twins correlate .9 and .5 respectively, the objection is made that the difference has little meaning because in fact the dizygotic twins have a wider range of environmental variance. It seems inconsistent to argue that there is a wide environmental variance between two dizygotic twins in the same family but a narrow variance between two related families.

Professor Layzer agreed that all these studies indicate that the closer the degree of genetic relationship for a given environmental difference, the closer the scores tend to be statistically. This indicates very clearly that there is a strong genetic component. The point at issue in estimating the heritability is not whether there is a strong genetic component but whether there is a significant genetic influence.

Professor Jencks noted that by talking about the ratio of one component of variance to another one tends to lose sight of the fact that the actual sizes of the variances are what is really important, and the sizes in these studies are quite disparate. In the one American study of monozygotic twins reared apart the variance is only about 60 percent of that of the general population. Hence there is a question about the variance that was not included in that sample. Professor Layzer pointed out that the correlation between unrelated individuals reared together is important because it is the one correlation that does not suffer from the genotype-environment covariance difficulty; presumably that kind of study is potentially capable of giving reliable heritability estimates, since it gives a direct estimate of the environmental contribution to variance.

Professor Jencks commented that while this approach avoids one problem it does not eliminate the tendency of adoption agencies to select environments that correlate with genetic predisposition. Professor Layzer agreed that one cannot eliminate all the systematic effects by this design, but it does eliminate one that theoretically is the most serious. He quoted Jencks's figures, corrected for unreliability and restriction of range: .17 for one study of ten cases; .29 for a study of 21 cases; .46 for a study of 93 cases; and .72 for a study of 41 cases. In accordance with Fisher's criterion the proper mean one arrives at by combining those and giving them weights in accordance with the number of cases is .5 ±.05; .5 is the lower limit from these studies, not taking into account large systematic effects which work in both directions on the environmental component of the variance, giving an upper limit of .5 for the broad heritability; insofar as this particular determination is concerned, the broad heritability could lie between zero and .5.

Dr. Eisenberg asked whether there are ways of doing an experiment that would provide more reliable data, or is the problem related to the explanation of the data. Professor Loehlin expressed the view that there is no way of conducting a study that will settle everyone's doubts; probably the most useful approach would be to analyze the results of as many varieties of studies with different potential sources of error as possible. If twin data were not available, if there were only adoption study data, one would have much less confidence in the results; but because the twin data involve different problems from those of adoption studies and still give the same order-of-magnitude kinds of results, the data are more persuasive. Professor Trivers pointed out that studies of half siblings, especially those related to the father, and comparisons of those related to the father and to the mother, are especially useful.

Professor Mayr pointed out that while a parent and child share half their genes, two siblings theoretically need not share any—they could share 98 percent or 2 percent—and this is often overlooked when in twin studies only those dizygotic twins who are very similar are used as the standard of dizygotic twins. Since they are likely to be genetically very similar in everything else, in addition to appearance, they should not be used as a random comparison with monozygotic twins. Professor Loehlin suggested that it is now becoming possible to make rough measurements of genetic similarity on the basis of blood groups, etc., and so it might be feasible to find twins who look alike but are quite different genetically.

Professor Trivers raised the question of the motivation for getting a precise estimate of the heritability of a trait. If it is granted on the basis of the evidence available that there are bound to be genetic and environmental components to variance in intelligence, however defined and measured, what is the utility of or motivation for getting a precise figure. If one knew in what situations estimates of heritability were going to be applied, one could evaluate the necessity for precise figures; if it is determined that precise estimates are necessary, then the questions raised about possibly significant measurement errors in methodologies of heritability analysis are relevant.

Professor Lindzey indicated that while in animal work (where heritability measures are used to predict response in artificial selection experiments) the precision of the numbers is important, in twin studies it is a matter of relative indifference. Professor Mayr agreed that for many purposes precise numbers are unnecessary. It is important to keep in mind that even if heritability measurements are worked out very precisely such measurements still would probably be immaterial in connection with populations in different environments. The objectionable usage of heritability measures is to treat a heritability measure developed in one kind of population as a physical constant and use that figure for totally different populations. Heritability within a population probably can constantly be refined by the various techniques that have been suggested but it still would not be applicable to cross-cultural comparisons.

Professor Harrison commented that for predictions with respect to human populations the precision of heritability measures is crucial. If one wants to predict the effects of social mobility, a reasonably good estimate of heritability is necessary; therefore, it is critical to know whether in fact the estimate is reliable. Dr. Davis, however, declared that with respect to social policy a precise estimate of the heritability of IQ within the American population—whether it is .8 or .4—is not particularly important. A sense of the order of magnitude seems to be the critical point in connection with a consideration of the social implications, a recognition that there is a considerable genetic component as well as an environmental one. Hence it is important not to assume, just because there is considerable measurement error, that the results of heritability studies are worthless.

In Professor Loehlin's view it is better not to express estimates of variance in terms of heritability coefficients but rather to talk about the several different components of variance and avoid ratios between the environmental and genetic components. A certain amount of the controversy in this area is caused by people with different biases adding the ambiguous elements to either the genetic or the environmental component, which is not particularly useful scientifically.

### Multivariate Approaches

Multivariate approaches represent some applications of these general notions that may answer some interesting questions, not as value charged as those related to estimating the genetic or the environmental components of particular traits. One can use very closely analogous methods—twin studies, adoptions—to divide the covariance between two traits into genetic and environmental or interaction/correlation components. Some of the questions raised about intelligence—whether it is a general ability or a number of specific abilities—can be attacked in this manner. For example, verbal ability and spatial ability will be correlated in any broad population, and one can ask whether the correlation between these two abilities is more or less highly heritable than either of the abilities themselves. If the former, that might support inheritance of a general ability, which is differentiated into special abilities by environmental experiences. If the latter, it might suggest very specific genetic components, which tend to become correlated over all because of environmental factors such as better schools, better home environments, and so on.

This is considerably less complicated methodologically because the weaknesses in the methods are not so significant; for example, while the defects of the twin method are troublesome in connection with estimating the absolute level of the genetic component, they create less serious problems for estimating the relative level for different traits. One is interested not in the absolute size but in the relative size—is the covariance less than the variance or is it greater? This approach requires large samples, because one is essentially making comparisons between correlations where there may only be small differences in the correlation.

Professor Loehlin noted that his studies in this area showed an interesting although frustrating result: the covariances were neither higher nor lower than the variances of the trait but were about the same. However, to date this approach has been used only on a limited scale, and in his view it might very profitably be used in connection with the question of whether the different abilities measured in an IQ test are biologically unrelated or whether in fact there are relationships of interest among them. For example, it might be that what is common to two different tests is the fact that the same genes influence performance on both of them; and that what produces the difference in performance is different experiences—one person might be better at one test than another because he has had one kind of experience rather than another.

Another possible outcome would be that in fact the genes that influence these two performances are quite different, but the performances are correlated in the population because they are both influenced by the same kinds of environmental factors like quality of schooling, etc. A third possibility is that one will not find any difference between the covariance and the variances, which is difficult to interpret; it might mean that whatever these two tests have in common is relatively influenced about as much by the genes and the environment as what is specific to the two tests.

Lloyd Humphreys commented that he has been studying cross-twin correlations versus within-twin correlations for a set of 40 Project Talent variables in an effort to determine whether their ratios are different for different types of tests, traditional aptitude tests and traditional high school achievement tests. No matter how he grouped the 40 tests he found no differences in heritability as a function of type of test, and he concluded that in terms of heritability there are no differences between achievement and aptitude tests when all of the subjects are exposed to a high school environment. Professor Layzer pointed out that that result is reminiscent of a result that Lerner refers to for dairy cattle: broad heritabilities as measured for a wide variety of traits turn out to be between .9 and unity; no matter what is measured, apparently, with split MZ pairs one gets a result very close to unity for broad heritability.

The significance of that finding is still not well understood by animal geneticists. It could indicate a very large contribution of nonadditive genetic variance; it could indicate hidden genotype-environment interaction; or it could be an unknown factor. In his view one has to be very careful in drawing inferences from heritability estimates, for in the case of dairy cattle, despite the high values of broad heritability, one gets very low values for narrow heritability—.25, for example, for milk yield. Professor Loehlin observed that one would expect large differences between broad and narrow heritability in dairy cattle (which would not apply to humans), since these cattle have in fact been strongly selected artificially for things like milk yield, rate of calf growth, etc. After intense selection one expects to find a large nonadditive genetic component and hence a large difference between broad and narrow heritability.

## THE NATURE AND MEASUREMENT OF INTELLIGENCE

### Definition of Intelligence

Professor Humphreys opened his presentation by stating the definition of intelligence that he found to be most useful: the total repertoire of responses that are considered intellectual available to an organism at a particular time. The kinds of items that psychologists have by consensus accepted as "intellectual" are those included in the standard tests of intelligence such as the Stanford-Binet and the Wechsler-Bellevue, which sample the repertoire of intellectual responses under controlled conditions. This particular approach has a long history in ability theory, dating back to the early part of this century. Spearman, Thurstone, Guilford, and others referred to the abilities included in intelligence as entities, and Spearman, analyzing their correlations, decided that these could be explained by a unitary "mental energy." However, shortly thereafter Thomson showed that those same correlations could be accounted for on the basis of random sampling of the stimulus-response bonds and the underlying neural mechanisms required for performance on a series of intellectual tests.

He indicated that there is a biological substrate for the development of this intellectual repertoire, and while the genetic component is a very important part of the biological substrate, it is difficult on the basis of available data to estimate it except within very wide limits. Professor Humphreys suggested that the limits are between .2 and .8 but emphasized that this view is as much based upon a belief in the evolutionary continuity of a human being from lower animals as it is on the basis of family relationship data. In addition to the confounding of genetic and environmental influences in the correlations derived from family relationship data, the correlational data themselves are highly fallible. As Jencks has pointed out, depending upon one's source of data, one can arrive at considerably different estimates of heritability.

Also, when Jencks and others reanalyzed, using weighted means, the family relationship data reported by Erlenmeyer-Kimling and Jarvik (who used median correlations), in several critical cases the results differed substantially, and in the direction that one would not expect on the basis of the known sampling characteristics of correlation of data. For example, the reported median correlation between unrelated persons reared together was .20, while the weighted mean of these various studies was .35; that gives a quite different estimate of heritability. There is also a psychosocial substrate for intelligence; while many environmentalists assume that the psychosocial substrate can be estimated from a test score, it too is highly complex and, in Professor Humphreys' view, equally impossible to estimate with accuracy in terms of present data.

Professor Griliches commented that weighted means are not necessarily the best approach; medians are less likely to be subject to the influence of extreme studies. Since arithmetic means are heavily affected by extreme observations, there is a good argument for the use of medians. Professor Layzer felt that it is necessary to weight the studies by number of cases, unless one wants to make a value judgment; but value judgments must be made as to the accuracy of the study, circumstances of the way the data were collected, and so forth. In response to his query as to whether that was done in the studies referred to, Professor Humphreys replied that both studies took the data as given and simply summarized it.

Professor Feldstein questioned the quantitative importance of the point regarding differences in medians and means; it may make a difference in some individual cases, but taking into account the fact that there are other familial relations—each of which is an attempt to estimate the same parameter—individual differences may not significantly affect the overall estimate of heritability when the entire body of data is pooled. Dr. Davis commented that small differences between different studies are not surprising; since heritability is the property of a specific population in a specific set of environments, there is no reason to expect that different studies, even if precise, would give the same heritability.

Commenting that IQ tests test a great number of talents, Professor Thomson asked whether there is any reason to assume that there is one biological mechanism behind all these various abilities, whether the traits that compose the compound trait that is being measured are linked together in some way. Professor Humphreys replied that the assumption is that there is a very heterogeneous biological substrate that includes all of the structures of the central nervous system. The compound trait was developed from a mass of data that relate to each other in a meaningful way; a consensus arose from empirical observations of relationships.

Professor Humphreys called attention to some problems in connection with the definition of intelligence indicated previously. First, it implies that a test designed to be culture-fair would not be a test of intelligence because it would not sample the same intellectual repertoire as the Stanford-Binet or the Wechsler-Bellevue. It also suggests, for the present period of time at least, that the definition of general intelligence may not be general enough because it somewhat slights mechanical concrete problem solving; one can characterize the content of the two standard tests as being primarily symbolic abstract problem solving, and it probably should be somewhat broadened.

### Nature of the Intelligence Test

To clarify the nature and emphasize the limitations and assets of a psychological test, Professor Humphreys suggested a simple analogy. If one wanted to measure height in terms of behavior or performance rather than by a linear scale, he might do so by setting up a number of "L" shaped pieces with the horizontal bar at the top and the vertical bar varying in height. A group of men are measured: each one of these uprights is passed across their heads and the number of times the tip of the horizontal bar touches the scalp is recorded. A person with a low score is short and a person with a high score is tall. If the uprights are split into two random samples and separate scores are taken, the correlation between the two scores will be very high—the person who was tall on Test A would be tall on Test B and vice versa. Furthermore, if the combination of pass-fail scores is correlated with height, in the original linear scale of measurement, the correlation is very high. The behavioral test of height is cumbersome and time consuming compared with direct measurement; however, with highly standardized conditions there is no problem with respect to the rank ordering of individuals.

A second analogy is more closely related to what has been done in the construction of intelligence tests. If one wanted to measure height but had no physical measure and insufficient hypotheses to develop the standards used in the first example, he might try to measure height by getting a variety of indications of height from the person by measuring the length of the fingers, various dimensions of the palm of the hand, the length of the bones in the arms and legs, the depth of the trunk, etc., and adding these bits of information together in such a fashion as to get a reliable and valid measure of height. Given fifty or so different physical measures of these parts of the body, with the proper weights it would most likely be possible to form a scale that would correlate very highly with a direct measure of height. With height there is the outside criterion of weighting—one knows that he ought to give more weight to the long bones of the body—but with intelligence that property is not available. Various scholars in the past have assigned different weights to different segments of indicants of overall intelligence and have come out with different answers, but such measures are being gradually refined. In Professor Humphreys' view tests will be developed that will be useful measures of the overall construct, even though with each item one is measuring only a small piece of the ultimate goal.

Professor Humphreys then commented briefly on some of the characteristics of ordinal scales of measurement. Intelligence and other psychological tests are ordinal scales since they provide only rank order information. One can get highly reliable, consistent measures; but psychologists usually think of reliability in terms of correlations rather than of numerical errors of measurement, because the units of measurement are arbitrary and there is no point in talking about the error in the IQ in terms of arbitrary units of measurement. There is no problem with respect to validity; one can derive valid results from different performances, and the generality that comes out in the total score is not present in any one of the parts. The units of measurement are arbitrary and have no real meaning, and in different parts of the scale they may be different from each other; but this can be taken into account when doing research. The defects of an ordinal scale are not nearly as great as many people have supposed.

Members of the group asked about the operational meaning of reliability and validity. Professor Humphreys replied that the reliability coefficient is the correlation between two parallel forms of the same measure, while validity is the correlation between the measure and some outside criterion—academic success, performance on the job, and so on.

### Generalizations About Intelligence

Professor Humphreys proceeded to discuss some of the empirical generalizations that can be made about intelligence. First, it is a very robust trait. It is fairly easy to measure, with reliable, consistent results. It is not very much affected by environmental factors—temperature, humidity, fever, etc. It is also relatively robust with respect to the person who administers the test; there is some speculation on this matter, but the best data indicate the sex, race, and other individual differences make little difference. The trait is also relatively immune to coaching and to test experience; there is only a very slight gain from taking the same test in fairly quick succession.

Dr. Kety questioned the statement that variables of test administration—repeat tests, male-female interaction, etc.—make relatively little difference, and Professor Layzer pointed out that Jensen has reported that he increased the IQs of black slum children that he tested by talking with them beforehand and making them feel at home in the test circumstances. Professor Humphreys replied that this is a part of the basic standardization procedure; one would get highly variable results if he tried to give a group test to preschoolers, but under conditions where the test is appropriate and where instructions are followed, the variables cited do not make very much difference.

Professor Humphreys also called attention to several different correlations with respect to intelligence test performance. There is a correlation of about .5 between a parent and child in measures of intelligence; in the full range of talent the correlation between the socioeconomic status of the household, depending chiefly upon the principal wage earner, and the intelligence test performance of the child is about .4. This indicates that scores on intelligence tests are not a simple reflection of privilege. The correlation between family income alone and the child's score on an intelligence test is about .3. If the range of talent is restricted to applicants for college admission, the correlation between family income and academic aptitude is about .15.

With regard to outside measures of intelligence, within the school system scores on intelligence tests are most highly correlated with achievement in areas such as reading, arithmetic, and comprehension of oral and written material; correlations between intelligence test performance and spelling tend to be substantially lower. The occupational correlations seem to be strongest with those occupations that require performance most like going to school—the academic professions, law, medicine, etc.; there are relatively small correlations between scores on intelligence tests and other kinds of occupational success. Correlations with athletic success, selling occupations, musical talent, leadership qualities are not very high. Mechanical and technical occupations, other than the highest level of engineering, are not as well correlated with performance on intelligence tests as one would like, considering the importance of mechanical problem solving in a highly technological society.

Dr. Kety pointed out that it is rather arbitrary in the IQ tests how much verbal component and how much mechanical component one introduces, so that if one finds that people with mechanical ability do not rank very high on the IQ test, that may simply mean that the measure of IQ has arbitrarily emphasized the verbal qualities more than the mathematical. Professor Humphreys agreed and indicated that this problem was the reason for his earlier statement that the concept of intelligence should be broadened to include more concrete mechanical problem solving in addition to the abstract symbolic problem solving. Professor Lindzey observed that the emphasis on verbal qualities is not totally arbitrary because the origin of the test was to predict a particular kind of performance, which is not mechanical performance.

Professor Humphreys pointed out the technical difficulties with respect to obtaining high correlations within a given professional group. Of the brightest high school graduates almost 100 percent go to college; only a fraction of those further down the scale go to college. Of these, a smaller proportion graduate from college; and those intellectually mediocre individuals who do graduate have compensatory qualities of motivation, etc., that are very important in connection with occupational success. Therefore, low correlations among individuals within an occupation are not critical in determining the validity of intelligence tests. He also commented that there is not very much difference between achievement tests and aptitude or intelligence tests. If one obtains a composite in the sixth grade of all the achievement tests given in that grade and correlates that score with the score on the Stanford-Binet, the correlation is about as high as that derived from all the forms of the intelligence test itself.

Professor Humphreys then observed that with the indicated definition of intelligence it is not surprising to find that group performance has changed over a period of time. There is good evidence that there was a mean gain in the United States between World War I and World War II, and in Scotland between the middle 1930s and the late 1940s. As one probable factor, in this country about 30 to 40 percent more 18-year-olds were graduated from high school between World War I and World War II than before World War I. There is also a possible genetic explanation for the gain, that it could in part be the result of hybrid vigor; if any country in the world might illustrate hybrid vigor it would be the United States with its heterogeneous immigration.

Professor Layzer asked how scores on intelligence tests support a gain in intelligence if the mean is conventionally set at the same value all the time. Professor Humphreys replied that it is not set at the conventional value, that it is well within the state of the art, and that these data are based upon that art, to maintain that arbitrary zero at the same place in successive editions of the test and keep the arbitrary units at about the same level in successive editions of the test. It is thus possible to tie in present tests with older tests and determine whether there has been an increase in means. Between World War I and World War II the mean gain was almost a full standard deviation—about 15 points. Professor Jencks suggested that the magnitude of the gain depends on the test. An order of magnitude of 15 points can be derived from military tests but a comparison of the 1960 and 1937 standardizations of the Binet does not indicate such a dramatic difference.

Professor Harrison raised a question regarding the specific validity of individual questions on intelligence tests. He asked what was the rationale, for example, for introducing something like digit symbol, which was supposed to standardize for sex but is one of the components in which women regularly perform better than men. Professor Humphreys replied that ex post facto data show that such components do predict subsequent academic performance more or less the same as the other subtests. Professor Layzer asked for clarification on this point: since there are very reliable tests for testing various cognitive skills that are relevant to specific tasks in everyday life, like the ability to solve problems in arithmetic or to read prose of different kinds and understand what it means, why do psychologists bother with these indirect tests that have to be validated through correlations?

Professor Humphreys answered that all these highly specific skills have a very high component of specific knowledge whose presence or absence in the experience of the person is more or less accidental, and in general it has been found that better results are obtained by adding together items of different types. This brings out psychometrically what the items have in common with each other, and it dissipates some of the highly specific elements that enter into success or failure on a particular item. Professor Layzer suggested that it would be better to take the weighted means of specific tests. Professor Lindzey replied that such an approach was adopted initially by Cattell and others: if one wants to predict a performance a reasonable approach would be to measure a specific factor or variable or operation that is closely related to that. However, it turned out that those things did not predict as well as the complex achievement variables that Binet and Simon put together.

Professor Humphreys then went on to discuss the stability of individual differences in intelligence over a period of time. He maintained that the stability of intelligence test performance during the period of childhood and adolescence has been considerably exaggerated by many scholars. To indicate that individual differences in the intellectual repertoire do change over time, Professor Humphreys referred to data on scores on an intelligence test given annually between the ages of 8 and 18: the correlations between adjacent years tend to be high, while the correlations over the period of 8 to 18 years are a good deal lower. There is thus a good deal of change in the rank order of individual differences over a period of time.

While unwilling to explain this instability on either genetic or environmental grounds, Professor Humphreys emphasized the importance of knowing that a general intellectual measure, which is correlated with a great many socially useful things, is not stable during the early childhood and adolescent period. A University of Illinois study shows that the correlation between high school rank in class and first semester freshman grades is about .4, but this drops off to about .22 in the eighth semester of the senior year, so the instability of intellectual performance continues into the college period.

Professor Feldstein suggested that people may be selecting their own tests, that they are choosing courses in which they will do well so this need not have any correlation with innate ability. Professor Humphreys noted that it is also true that some people in easy curricula still do poorly; otherwise there would not be the constancy of standard deviations, the constancy of correlation between adjacent semesters. He went on to indicate that data for the graduate record examination given in the senior year show that GRE verbal and quantitative scores are most highly correlated with freshman grades and least highly correlated with senior grades. The correlations for the advanced test in the subject matter show a different pattern, with the highest correlations being in the sophomore and junior years. The advanced test, across disciplines, country wide, gives better predictions of graduate school performance than do the verbal or quantitative ability tests. Professor Thomson commented that the GRE is a kind of minimal intelligence test but beyond that it is not predictive at all. Professor Humphreys objected that while the slope of the regression line is not very great, with a correlation of about .3 depending on the field and the test used, the correlations with graduate grades are not zero. All the correlations are low, but it must be remembered that the range of talent of students in any one graduate school is very restricted.

Professor Humphreys cautioned against drawing conclusions on the basis of small samples from a single institution. Data nationwide show that the quantitative test does provide graduate criteria in physics and engineering and the verbal test provides graduate criteria in English and political science and history; but again nationwide and across all disciplines the advanced test gives somewhat higher predictions than either the verbal or the quantitative, and the advanced test is not so predictive as undergraduate grades. Professor Jencks pointed out that the predictive validity of these scores varies from institution to institution. There are enormous variations across institutions regarding how well the SAT does, for example, in predicting performance, which presumably has something to do with what kinds of performance different institutions ask for.

In a final comment, Professor Humphreys called attention to a researchable hypothesis of long standing that has never been tried out adequately: that with growth, maturation, and experience—the mixture of heredity and environmental influences—there is differentiation of human abilities. Professor Humphreys expressed the belief that this differentiation cannot take place at any level of ability, that there must be some given minimum of general intelligence (at least general academic intelligence) before one can start branching out into law, chemistry, engineering, etc., and that rather than capitalizing on existing talents, people in undergraduate work in college are in reality developing narrower, more specific ability out of this general ability that is defined as intelligence. He suggested that this is primarily an environmental development. However, Dr. Davis felt, from general observations of student behavior, that this differentiation does not result primarily from an environmental factor: the student realizes, on the basis of performance, that he has more talent in one direction than another, and he directs his interests accordingly.

### Other Human Behavioral Traits

Dr. Davis pointed out that the general question of human diversity and its origins involves many important behavioral traits. There is a tendency to focus on IQ simply because there is much more information on it, the numbers can be manipulated mathematically, and the results have practical value in predicting educational achievement. But such excessive emphasis can be dangerous in its social implications. From the point of view of the biology of the nervous system and its influence on behavior, the IQ test is not a sharp research tool; it is a practical test that has been developed for the purpose of guiding educators. If one wanted to analyze the intellectual operations of the brain one would go along the lines of fractionating the whole into more specific, smaller components of the overall mental processes.

While work along these lines has not been of particular value in terms of its guidance for educators, it will almost certainly become of increasing value in future studies of genetics of behavior, in the same way that all fields of science become more analytical as they acquire more information. Also, the intelligence test is very often a mixture of tests for abilities and for learned knowledge. Because these correlate so well, that mixture may be satisfactory from the point of view of the practical needs of education, but from the point of view of the biology and genetics of behavior that kind of mixing together of what has been learned (phenotype) and something closer to the capacities of the individual (genotype) is unsatisfactory.

While the definition of intelligence as a particular repertoire of intellectual activities may apply accurately to this kind of test, a more comprehensive definition is needed that will also apply to other complex mental operations. Dr. Davis suggested the following: the ability to assimilate information, to recognize and create complex relations between its components, and to communicate the results. Such a definition could be applied not only to abilities for abstract reasoning, which are emphasized in the IQ test, but also to the complex processes reflected in social forms of intelligence, which are harder to quantify but which are very important in relation to success in business and practical life; this definition would also apply to artistic and musical intelligence. These are very important aspects of the nervous system and its functions, and they should be included in an analysis of human diversity.

An implicit assumption has prevailed for a long time that a high IQ is the most desirable trait that a person can possess and therefore it reflects some special human worth. Perhaps some of the controversy surrounding the arguments about how IQ is distributed could be dissipated if it were emphasized that while IQ is of biological interest, and it has educational and vocational value in terms of predicting what one can and cannot do well, it is not in any sense a measure of human worth.

Dr. Eisenberg commented that the performances that can be measured with the greatest degree of precision and replicability are the least interesting as far as carrying any implication for complex performance. IQ tests predict school performance, but they are not very successful in predicting postschool behavior. The differences between people's later performance in life after they pass the educational screen are rather poorly predicted by any measure of performance in college, from which one could draw the conclusion that one does not know how to assess those traits that are involved in the complex tasks that life calls on people to perform.

Even if IQ is highly heritable, and even if it were possible to skew the population toward developing more and more people who could perform well on IQ tests, that might not be desirable because it might inhibit the development of those other attributes, not now measured, which are in fact very important. Professor Mayr added that personality factors greatly influence what is considered success in life; studies indicate that one's degree of self-confidence plays an important role in relation to occupational success.

## AGENDA FOR FUTURE MEETINGS

There was considerable discussion about the future course of this series of meetings. Dr. Davis indicated that it had been planned to devote the next session to group differences but in view of the lack of agreement on the reliability of heritability estimates this might not be profitable. It might be preferable to go on to a discussion of social policy implications on the general assumption that there is substantial genetic variation within the species and a good probability of significant genetic variation between groups in the species without trying to attach a number to the genetic component. However, Professor Heymann expressed the view that it would be a serious mistake not to address the crucial argument that one can draw from heritability figures some conclusions about how much of the variation between races is genetic and how much is environmental. Since this involves a body of ideas that affects people's self-image and their image of other people, it is an important social phenomenon. By drawing on the genetic and evolutionary evidence and methodological data presented at these seminars, this group might be able to determine with some confidence whether the argument that is made from heritability figures to a high heritability of interracial differences is very weak, very strong, or something intermediate.

Professor Jencks felt that there was a need for further evidence relating to the heritability of individual differences. Indicating that it was his impression that the problem of IQ heritability was one of the major reasons for convening this series of meetings, Professor Trivers also felt that further analysis of the data relating to the heritability of IQ was needed; he asked for further information on the data that have been gathered and the methods by which they have been gathered. Those who support the view that there is a high heritability of interracial differences claim that there is a wide literature that leads to certain strong conclusions in this direction; this information should be presented to the group before the discussion proceeds to the implications of group differences. Professor Griliches supported the view that before going on there should be further discussion of the empirical quality of the evidence and the quality of the analysis of the evidence. Professor DeVore noted that the best evidence for theorizing about group differences comes from studies of monozygotic twins reared apart, but the group has not discussed the methodological problems in any detail. He also felt that in a discussion of human diversity further consideration should be given to sex differences.

Dr. Kety commented that another body of knowledge that should not be overlooked is the environmental causes of human diversity. He expressed the view that the body of evidence relating to the environmental effects on IQ, as far back as the in utero environment and early life experience, should be examined as critically as the evidence relating to genetic factors. There is a considerable body of data on the effects of prematurity, nutrition, etc., on the development of the brain. Professor DeVore suggested that the data relating to cross-cultural comparisons would be of interest in relation to a discussion of the environmental influences on intellectual functions.

Dr. Eisenberg observed that Jensen certainly believes that the available evidence warrants the kinds of conclusions that he draws; others are not persuaded that those conclusions are legitimate. An important question is to determine how one goes about setting forth in a way that is scientifically legitimate the facts that would be required to support the view that one can extrapolate from within-group heritability estimates to across-group differences. Professor Feldstein noted that no matter how much evidence is accumulated and analyzed, some will feel that one cannot rule out a conceivable unknown factor that has depressing effects on intellectual development, and they will never be satisfied that any conclusion is valid.

Professor Brooks suggested a discussion of what kinds of data would cause one to try to change social policy. Rather than a concern about what the data are, we might assume, for example, an average IQ 100 percent heritable difference of 15 points between two groups and try to determine what would be an appropriate social policy in response to that. Dr. Davis agreed that it would be useful to make several different assumptions and explore their implications; there are certainly areas of public policy where it would make a difference whether the heritability of a trait is negligible or substantial.

Professor Thomson expressed the view that with respect to public policy it may not be relevant whether IQ is highly heritable or not. It would affect the reasons one would give for establishing certain policies—if it is shown that IQ is highly heritable the opposite assumption could not be used to support a policy of reverse discrimination—but in her view the question of heritability may not have direct bearing on whether such policies should be implemented. Professor Mayr pointed out that for some questions the heritability of IQ would be of direct interest: for instance, if the U.S. adopts a policy of zero population growth and if it were shown that undoubtedly there is a high genetic component in intelligence, it might very well be a matter of public policy whether families who can be predicted to produce children with a high intelligence should have more children than others.

Dr. Davis cited a policy issue relating to education: if we proceed on the assumption that children from all socioeconomic backgrounds should reach the same level of achievement in school, their failure to do so is attributed to inadequacies of individual teachers or of the educational systems; but if academic capacities are in fact highly heritable and are not evenly distributed among SES groups that assumption would result in unjust and destructive evaluations of teachers and educational systems. Therefore, whether or not there is a strong genetic component in intelligence does have a direct bearing on policies aimed at improving the educational system.

# Seminar IV: Intelligence and Group Differences (May 10-11, 1974)

## MAJOR PRESENTATIONS

Lee J. Cronbach: *The Nature and Limits of Intelligence Testing*

Arthur R. Jensen, Luigi Cavalli-Sforza: *Racial Differences*

Christopher Jencks, Geoffrey A. Harrison, Richard J. Herrnstein: *Socioeconomic Class Differences*

*Chairman:* Leon Eisenberg

## RECAPITULATION OF SEMINAR III

Dr. Davis called attention to several points that seemed worthy of brief reinforcement. In the research on experimental animals reviewed by Professor McClearn, genotypic effects on behavior could be rigorously demonstrated because it was possible to breed genetically homogeneous strains and to make the environment quite constant: hence the differences seen were almost entirely genotypic in origin. This evidence for large genotypic components in animal behavior supports the expectation of similar processes in man. Moreover, strains that have been inbred with respect to nonbehavioral traits often also show substantial differences in their behavioral traits. That observation is relevant to the general argument that groups that have been separated for a long time are likely to have differences in behavioral traits even if these have not been selected for.

Dr. Davis also observed that because of the highly polygenic nature of behavior it is difficult to discuss its genetic basis except in very general terms. However, one can in fact estimate the number of genes involved in one kind of behavioral defect: hereditary deafness, a recessive disease which appears when an individual is defective in both copies of a given gene. Individuals with this disease generally marry each other, and their children are ordinarily not deaf. In such cases the parents must be blocked in different genes, so the defective allele from either parent is complemented by a normal allele from the other. However, an occasional pair of deaf people regularly produce deaf progeny, which indicates that they are blocked in the same gene; and the frequency of such pairs indicates that there are at least 100 different genes that can cause hereditary deafness. This is perhaps a minimal estimate from which to start when trying to guess at the number of genes involved in other, more complex aspects of behavior.

Finally, Dr. Davis discussed the finding, reported at the last meeting, that if inbred strains were exposed to as constant environments as possible experimentally, the residual range of variance, with respect to expression of the trait being analyzed, differed markedly from one inbred strain to another. Since the extent of the response to uncontrolled variations in the environment thus varies from one genotype to another, the comment was made that this is a clear example of the importance of gene-environment interaction. This statement is correct for the circumstances of the experiment; but Dr. Davis questioned whether the finding has any relevance for the importance of interaction under ordinary circumstances. For if one deliberately makes the additive genetic component as small as possible by using inbred strains, and if one makes the additive environmental component as small as possible by making the environment as constant as possible experimentally, then by definition what is left is interaction. Hence we were discussing a special case of interaction made to predominate experimentally, rather than an example that can be extrapolated to ordinary circumstances.

## THE NATURE AND LIMITS OF INTELLIGENCE TESTING

At the previous meeting the issues relating to the nature and measurement of intelligence were outlined by Professor Humphreys. However, in view of the importance of this subject it was felt that another session of the seminar should be devoted to this topic. As background for the continuing discussion Professor Cronbach prepared a comprehensive memorandum discussing the major questions relating to intelligence tests (see Appendix I).

Since it had been suggested that it would be useful to the seminar to have a more concrete and operational notion of what typical mental tests consist of, Professor Cronbach first described several specific items from the Stanford-Binet. He showed examples of tests at an early level (2½ years) that demonstrate the child's ability to understand the questions and his acquaintance with simple concepts—identifying objects for use ("show me what we use to iron clothes"), identifying parts of the body ("show me the doll's hair"), naming objects, picture recognition, etc. In tests at the Year 7, the emphasis shifts from everyday simple performances to predominantly verbal tests. These include picture absurdities, relationships (in what way are apples and peaches alike?), reproducing simple shapes, etc.

One of these subtests includes the question, "What should you do if you find a three-year-old baby on the street who was lost from its parents?" This type of question is particularly likely to be challenged as having a cultural loading—the right response for a person in a middle class suburban neighborhood may not be the most intelligent response in the inner city. At a higher level, opposite analogies are utilized ("rabbits' ears are long; rats' ears are ?"), as well as number series. At the adult level there are tests of straightforward reasoning (water jar problems) and of the relatively subtle use of vocabulary (what is the difference between laziness and idleness?). Professor Cronbach indicated that these examples show that these tests are not heavily loaded educationally.

Professor Cronbach then turned to the Lorge-Thorndike test, in which a critical point is the distinction between the verbal and the nonverbal sections; the verbal section is what conventionally marks IQ for the schools. The verbal and nonverbal are substantially but not perfectly correlated; each points to a somewhat different kind of individual difference, and the verbal test consistently is a better predictor of school work. For example, in the ninth grade the Lorge-Thorndike verbal correlates .60 with later high school grades, while the nonverbal correlates about .40. The difference is not always as great in the earlier grades (it varies from 10 to 20 points), but there is a consistent difference in prediction. This has led schools to deemphasize the nonverbal test. However, the intelligence test of the verbal sort yields little information beyond what the school is already learning from achievement tests; it really has little function except that it is curriculum-free.

It is difficult to design an achievement test that is not conditioned by the kind of mathematics that has been taught, or by the sources selected for reading, social studies, and the like; hence any single achievement test may be an unfair basis for comparing persons from different schools. One of the questions that should be considered is whether an effort should be made to reverse the present emphasis on the verbal test, in order to capitalize on the potential of nonverbal types of tests to identify those who have a kind of intellectual excellence that does not correspond very well with the school in its present form.

Professor Mayr inquired whether any correlation has been attempted between nonverbal tests and individual school subjects rather than overall school performance. Professor Cronbach called attention to data which indicate that school grades in science are predicted better by the verbal-numerical combination, math by the numerical, English by the verbal-numerical. (Although the numerical correlation in the latter might seem odd, this is a fairly common experience; people who do well in the verbal school subjects also do well in the mathematical.) Only in shop did the Abstract Reasoning test (which corresponds to the Lorge-Thorndike nonverbal) predict as well as the others. Dr. Davis asked whether the lack of correlation between the abstract test and achievement in science indicates that science courses require more "spitting back" and less reasoning than one might hope for. Professor Cronbach emphasized that the verbal tests are also tests of reasoning; he felt that it would be hard to argue that reasoning about drawings, which requires analogical thinking and use of such concepts as proportion, is any more or less scientific reasoning than reasoning about verbal analogies.

Professor Griliches asked for further elaboration about the basis for believing that the nonverbal test, even though it does not correlate very well with school performance, still measures something important. Professor Cronbach indicated that when a child shows good performance on tasks that are unquestionably intellectual, but is below average in school work, it is conceivable that one could design an educational treatment that is more fruitful than conventional schooling. He referred to the 1940 experiment of G.L. Anderson, described in his memorandum, which indicated that fourth graders who had high IQs and poor pretest scores in arithmetic profited from meaningful teaching in the next grade, where the students were encouraged to think through a relationship and work out procedures for themselves; the students with good initial skill in arithmetic but comparatively low IQs did their best when drill was the principal teaching method.[a]

This is referred to as aptitude-treatment interaction. For a particular aptitude measure there are presumably some treatments that have a steep regression on the measure and other treatments where the slope is shallow. Although this line of research has been a frustrating one, it is hoped that, as different kinds of intellectual ability are differentiated by mental tests, new modes of educational treatment can be designed to facilitate performance. Clearly in educational methodology the need is for diversity of opportunity (to quote Jensen) rather than simple equality of opportunity.

Dr. Eisenberg commented that one of the problems in getting data that correspond with expectation relates to teacher behavior. Teachers may say and truly feel that they are producing an enriched environment, but without adequate observation of performance in the classroom one cannot be sure that the teaching environment is significantly different from the original; hence a new method may appear to be unsuccessful even though the prediction is real. Professor Cronbach agreed that the teacher is an important source of variance; as studies in this area proceed, it may well be found that some of the important interaction relates to the teacher-to-teacher variation rather than to the utilization of different educational treatments.

Dr. Davis noted that one generally thinks of abstract reasoning as being central to intelligence, yet it correlates less well than verbal and

[a]While this line of investigation is important, the Anderson conclusion appears not to be sound. Cronbach and Webb have recently reanalyzed the data, separating between-class effects from within-class effects. They find no trace of interaction at the within-class level. The large effect that appeared between classes, which accounted for Anderson's original "finding" from the pooled data, can be attributed to an accident of sampling.

numerical reasoning with school performance. He asked whether there have been any studies to determine whether it correlates better with some aspect of subsequent performance in life. Professor Cronbach indicated that he knew of no data showing that abstract reasoning correlated better than verbal reasoning with any kind of performance. Arthur Jensen commented that both kinds of measures have a large *g* component, so there is something common to both kinds of tests that is neither specifically verbal nor specifically abstract; to designate the nonverbal test as abstract, as if the abstraction ability is a primary, such as verbal or numerical or spatial, is a misnomer.

Jerome Kagan also objected to the distinction between abstract reasoning and nonabstract reasoning; he suggested that the differentiation be expressed in terms of reasoning with verbally presented concepts, reasoning with pictures, reasoning with designs, etc. Professor Cronbach noted some confusion between the notion of "abstract" reasoning, which may occur in many kinds of test, and the particular figured test labelled Abstract Reasoning. In psychologists' references to abstract or nonverbal reasoning—the current term is "fluid ability"—the emphasis is on adaptation to a new problem. However, verbal encoding is undoubtedly involved. While these tests are nonverbal—no word need be spoken, they can be given in pantomime—it is inconceivable that the reasoning does not have a large symbolic verbal element.

Dr. Davis suggested that abstract reasoning, in the sense that it is tested for, may relate to the distinction made by philosophers of science between deduction and induction. While these terms are not fashionable today, in his book on *Race*, Baker uses the concept of induction quite frequently as a taxonomist; he speaks of some people as having a talent for induction, by which he means identifying common features in a large number of patterns. This may be the central feature of the test of abstract reasoning—the ability to generalize, to find a common feature in somewhat different patterns—and perhaps that kind of reasoning does not enter to any significant extent into most academic or vocational situations.

Professor Griliches referred to the philosophical scientific question regarding the construction of scales. In addition to those that measure IQ there are a variety of tests to determine such things as fear of failure, achievement motivation, etc., in which people are given a set of questions, they are scored, and an index is computed. The question is how to distinguish a good test from a bad one. Indicating that the bulk of the literature refers to an internal reworking of machinery that has not yielded significantly useful results, Professor Griliches asked for further comment on the question of test validity, especially since the figures derived from IQ tests are used so frequently in connection with the heritability issue.

Professor Cronbach agreed that each psychologist and sociologist tends to devote excessive attention to his own experiments. He may correlate them with outside variables, but few people force an instrument into a larger context. However, the Rotter locus-of-control scale—which deals with how much a person feels he has control of his own fate, how much is success a matter of luck or the will of the gods—is consistent with observations of various cultures made by sociologists and anthropologists using methods quite different from those of psychologists. The interest in this scale comes not merely from the fact of a single psychologist's developing a gadget but from the fact that observations of very different types converge.

Psychologists in the ability field have emphasized predictive correlations as indications of validity, and this is obviously relevant if one is merely trying to select students or employees. Beyond that, one can frame hypotheses that have a certain plausibility—for example, that it might be possible to break the predictive validity of verbal ability for a particular job if the training is modified in a certain way. This work is a comparatively recent development, and the testing of such hypotheses has tended to be done on a small scale; but there is a movement away from the simple question—what does it predict in the real world?—toward trying to determine the processes in which intervention would be useful. It is much more likely that one can intervene successfully to strengthen a specific process than to strengthen some global aptitude.

As far as the heritability issue is concerned, it is a historical accident that this is associated with IQ and not with general achievement. What is known about the predictive qualities of intelligence tests, about group differences, about the correlates of IQ generally, is indistinguishable from what is known about achievement tests. The only additional point to be made is that there is an important mystery in the nonverbal tests, which look as if they represent a respectable type of intellectual functioning that social institutions are not now capitalizing on; but there is no reason for suggesting that the ratios of the variance components are different for the verbal and the nonverbal tests. Professor Mayr observed that the difference between the predictive value of verbal and nonverbal tests becomes important since the two are combined in the final IQ values, and the difference between the verbal and the nonverbal scales may be greater in some groups than in others. However, Professor Cronbach cited data which contradict that suggestion where the differences between the social classes were essentially the same in four types of tests.

On the other hand, recent studies indicate the possibility that spatial ability separates off in a sex-related manner from the more general verbal ability. Professor Jensen added that Bock and Kolakowski hypothesized an X-linked recessive gene that enhances spatial ability. Since spatial ability is the one primary mental ability that shows a significant sex difference in favor of males, this would be predicted by an X-linkage hypothesis. Quite convincing evidence was presented for a single recessive gene effect in terms of the patterns of correlation between father-son, father-daughter, mother-son, mother-daughter, and siblings. Professor Kagan questioned whether there really are sex differences of a genetic nature in spatial reasoning. He indicated that outside of western cultures the data do not show any sex difference, and in a study done in the Boston area no sex differences were found until grade seven. In his view sex differences in spatial reasoning are culturally acquired.

Professor Jensen maintained that that conclusion does not follow prima facie from the evidence. It may be that in some populations the single recessive gene that enhances spatial ability is lacking, or it may exist in all members; in either case there would be no sex difference. Professor Trivers asked about the evidence for the existence of a single gene located on the X chromosome. Professor Jensen indicated that the form of the evidence is in the pattern of correlations between the parents of the same and opposite sex and between siblings of the same and opposite sex. A genetic model would predict a different pattern of correlation for an X-linked gene than it would for an autosomal gene. Dr. Cavalli-Sforza commented that correlations are fairly weak evidence; a better way of testing the hypothesis would be to find linkage to a sex-linked gene. Such investigations are proceeding.

Regarding the validity question, Dr. Eisenberg commented that with a longer interval between the test and the outcome measure that one is concerned with, generally the correlations tend to fall off; if the prediction is accurate only for a very short term, one cannot have much faith in the robustness of the measure. For example, scores on achievement tests of those entering professional schools are poorly correlated with final scores and with professional success. Professor Cronbach indicated that the low correlations can be partly explained by restriction of range and also by shifting criteria. What makes one a successful lawyer or doctor has relatively little to do with the kinds of intellectual skills that are tested for. Moreover, the people who have made the greatest scientific and scholarly contributions to law and medicine are not the ones who are most successful in terms of an income criterion.

The heart of the issue is that one can make some valid predictions about the inability of a person to get through the present academic curriculum, but there has been no effort to develop a kind of training suitable for such people who may have sufficient intellectual ability to make good physicians and lawyers. Dr. Eisenberg pointed out that that is one of the real limitations of IQ tests, which is quite independent of the heritability issue. Very few people get through a law school or a medical school with an IQ below 100. But optimum performance in some aspects of these professions may in fact not require the abilities measured by IQ tests. For example, many medical corpsmen carry out many of the tasks of the doctor very well but most of those people cannot gain admission to medical schools because they do not have the IQ and the talent for mastering the academic requirements, even though they would likely become good general physicians. Dr. Davis noted that this discussion seemed to be concerned not with a limitation of IQ tests, but with the validity of the standards of performance set by various professional schools.

Professor Cronbach agreed that tests are designed to predict success in a culture where entry to a large number of high prestige activities requires the passing of hurdles set up by the academic profession. But there would still be a question of who might succeed even if the requirements were not academically oriented. To emphasize this point Professor Cronbach called attention to an Army study in which a large number of tests were applied to candidates for flight training, tests with a large verbal or mathematical component, spatial tests, coordination tests, etc. However, they predicted survival only with a correlation of about .40 in the groups who survived the testing screen when, in a special sample that passed ordinary medical screening but was not screened on the tests, the validity coefficient was .77. There is considerable evidence that vocational success is a diffuse matter depending on many things other than technical job proficiency. But the issue raised is an important one. Psychological tests predict within a system; they perform a conservative function by screening out the raw material that our present machinery cannot handle efficiently. The important issue is that we do not know what some other processing operation could do with this reject raw material.

Dr. Davis agreed with Dr. Eisenberg that if the pattern of the practice of medicine were to be changed so that the people who take care of the more routine matters are a separate group from those who handle more complex problems, then a very different kind of medical training would be required and people with an IQ of 90 to 100 might fulfill some of those jobs very well. But he questioned whether that is relevant to the question of whether these tests are useful as educational guides. Dr. Eisenberg indicated that these tests are certainly an efficient mode of predicting who will succeed or fail in the system as it now operates. He emphasized that it is not a question of "routine" vs. "complex" tasks. The nonbiological tasks that the doctor might perform—sympathy, interest, relief of discomfort, etc.—might be very complex, yet require skills other than those recorded in academic performance. If one knew enough about these other complex skills that directly affect performance, it is conceivable that one could design another route of training that would allow expression of talents other than those necessary for school success.

Dr. Davis defended his fractionation of medical care into complex problems and less complex or routine tasks. While social sensitivity is a kind of complexity, the kind of reasoning that IQ tests are concerned with—the capacity to handle complex mental processes—is very relevant to difficult diagnostic problems where many factors have to be taken into account. This requires analysis that would be beyond the capacity of a person with an IQ of 90. Dr. Eisenberg agreed that making judgments of that sort represents a complex activity of an intellectual nature. However, there may be another kind of complexity, in what are usually considered routine, day-to-day tasks, which present tests are not designed to identify and that nobody knows at the moment how to measure. Some people may be gifted with social skills that are not correlated with academic attributes.

Professor Cronbach commented in a more general way on the problems of identifying and measuring the skills involved in successful job performance. He indicated that the requirements that have been traditionally imposed for entry into professional schools have had large elements of arbitrary elimination. It is difficult to see any relevance of four years of Latin in terms of overlap with the eventual occupation. Latin was being used the way the scholastic aptitude test was to be used in a later period, as a means of weeding out people who did not have the combination of intellect, discipline, persistence, etc., to do prescribed academic tasks; these attributes are not irrelevant to demanding professional jobs. There are a large number of jobs, however, and some of them are in professions where one can perform on the basis of overlearning as distinct from adaptation. In accounting, for example, while in certain circumstances very high levels of judgment are required, most cases of the work of accounting can be done by following some rules and by being diligent. It is only at higher levels of responsibility that abstract intelligence is at a premium.

The mental test is an attempt to measure adaptation, and many academic tests as well are designed to test the student's ability to deal with problems he has not seen before, his ability to transfer his knowledge. This issue of transfer—using knowledge in a new context in the face of additional complexities—is the kind of intellectual ability required for performance at the upper level. Unfortunately, it has not been possible to separate this highest level from academic competence because the person who has this ability also masters the lessons; but there is a significant distinction between exercising high level, overlearned skills in a familiar context and analyzing with reasoning.

There is also a difference between success at the high professional level where the analysis is essentially intellectual and the success that is essentially one of personal salesmanship. With regard to social skills, a promising area not yet exploited is group performance. If good tests of group performance were to be developed, they might indicate an intellectual function of great importance relating to social interactions. Dealing with other people—vis-à-vis solitary problem solving—requires a different kind of intellectual performance, a different manner of work, a different set of motivations, and so forth.

On the question of validity, Professor Kagan noted the assumption that children who get high scores on verbal analogical items and vocabulary items learn more quickly than others. He suggested that it is necessary to analyze how representative these items are in the experience of children in the U.S. If they are nonrepresentative, then high scores may be as much a function of differential exposure as they are of intellectual ability. Professor Cronbach commented that there are tests of different sorts. The Binet list is graded and the words at the adolescent level certainly are differentially distributed throughout the culture, whereas other tests use common words available to everyone. But the fact is that tests on simple words behave statistically like tests with relatively rare words.

A question of greater concern relates to vocabulary in the curriculum. One of the current issues is whether alternative language systems like Black English have the necessary precision for discussing the concepts in the standard curriculum. If there are equally good alternative language systems, this is an important challenge for the curriculum. On the other hand, if these concepts do require an advanced vocabulary, the validity of the test cannot be called into question because it employs such a vocabulary. That the vocabulary in intelligence tests is not universal almost certainly is not the heart of the issue of fairness across cultures. If it were, there would be dramatic differences between blacks and whites on the verbal test (not because of race but because of social class and education of parents and other covariates), accompanied by a much less impressive difference, or perhaps none, on the nonverbal. This is not the case.

With regard to the question whether test items are fair to people of different cultures, Dr. Cavalli-Sforza suggested analyzing the purpose of intelligence tests. If the purpose is entirely an applied one of deciding who should go to what school, etc., it may be that it does not matter whether the test is culture fair or not, because it is quite possible that by the time the child is tested, whatever his genotype was, he has already been molded into a phenotype which is irreversible. While this conclusion is unpleasant, present social conditions are such that a person can have lived in an environment which has lowered his IQ irreversibly, and perhaps one has to accept that hard fact in placement decisions. This, of course, presumes that IQ tests can accurately predict performance in a given culture, which is not necessarily true.

An interesting approach to determine whether test items are fair to persons from different cultures is to test in extreme conditions, where it is easier to pick out relevant elements of culture. Primitive cultures may lack many elements that are important in understanding, for instance, a pictorial test, but we tacitly take for granted that this does not matter. In fact, people from primitive cultures cannot easily recognize photographs even of their relatives or of themselves if they are not used to seeing photographs. Geometric figures are extremely unfamiliar to many primitive people because they never drew them, they do not have paper and pencil, they may even not have sand to draw on with a stick. Even in our culture the environments of many children have not included much paper and pencil, and they will not be familiar with these things.

Professor Cronbach indicated that while cultural differences may be relevant in connection with the response to certain items in intelligence tests, the essence of the psychological data is that the complex of intellectual performance is broadly defined and includes both verbal and nonverbal analysis. While there is some specialization—some persons may perform well on a narrowly isolated task and badly on a general mental one—for the most part the relationships are much the same for verbal and nonverbal analysis.

Professor Layzer asked for comment on the results of two recent studies. The first, which relates to the question of specialization, involved tests of simple logical inference in several different contexts. The results indicated that the ability of the subject to arrive at the correct answer was greatly dependent on the context; while the experimenters expected to find that context would not greatly affect performance on these tests, the results showed striking context dependence. The second study bears on the question of the stability of IQ. McCall's report of the massive Fels longitudinal study showed a substantial systematic variation in IQ between childhood and adolescence, a variation of about 30 points on average, and one 74-point variation was found. These variations fall into systematic patterns, and in some cases there are strong environmental correlates to the patterns. Professor Jensen indicated that correlations between IQ at one age and another vary greatly; when the earlier measure came before age five they are quite low. He referred to a table in his book which showed that IQ at age one-year-nine-months correlates .07 with IQ at age 18; with such a correlation a variation even larger than 30 points would be common. At age 2 the correlations with age 18 is .31; at 3 it is .35; at 6 it is .61; at 14 it is .76.

On the matter of stability of IQ, Professor Cronbach reported that Professor Humphreys and he had engaged in a considerable interchange, because Professor Humphreys had presented a set of data in the previous seminar that in Professor Cronbach's view seemed to overdramatize the changes between ages 8 and 18. After examining various sets of data, the two agreed that the best estimate of the correlation between IQ at age 8 and IQ at age 18 (corrected for error of measurement) would be about .75-.77. This implies neither perfect stability nor haphazard variation. The late bloomer cases are not a mythical phenomenon; on the other hand there are many cases where the change from 8 to 18 is essentially zero. The change of 74 points in one case almost certainly represents a result from one of the infant tests, which measure something that has little relevance to later academic performance. With regard to measurements of logical inference in different contexts, Professor Cronbach expressed the view that change in motivation or orientation is not likely to immediately convert an inferior into an excellent performance, although in many cases such a change will have some effect.

On the stability question, Professor Herrnstein suggested that it might help to compare IQ with something like physical height, which also shows correlations of this order. The parallel suggests that mental growth has a kind of stability much like that in physical growth, both having a certain amount of static and irregularity. With regard to the question of test validity he pointed out that one of the things that testers have had in mind all along, although not necessarily in a formal way, is the question of heritability. While looking for instruments that predict scholastic and other success, they have also been deeply concerned with the notion of native intelligence, which is the old-fashioned way of referring to heritability.

Professor Jencks added that if one looks for stability, which has been a more explicit goal, he will certainly rule out many things that may have a high validity simply on the grounds that they are unstable. The relationship between stability and heritability deserves considerably more attention. If one concludes that the correlation between scores at age 8 and age 18 is .77, and if one assumes that the genetic factors which influence IQ at those ages are constant, it is difficult to construct a model in which the combined effect of the genetic variance and the common family background variance and whatever covariance there is between the two is more than .77. If that is true it represents another item to be entered into calculations of heritability. However, it is also reasonable to assume either that the genetic factors at age 8 and at age 18 are different, or that heritability at 18 is lower than at age 8 because the environments are more varied. We would thereby eliminate the difficulty of using stability as a maximum estimate of heritability plus family background; but we would then have to rethink the usual assumptions about a set of genetic influences that are stable over time.

Dr. Eisenberg called attention to a study of middle class children which indicated that about a quarter of the population showed an IQ increase from age 6 to age 12 of as much as 30 points; another quarter showed a decrease but of a considerably smaller amount. From the policy point of view, if one were to assume that the figure at age 6 was a firm predictor and streamed children accordingly, he would do an injustice to a not inconsiderable population, since streaming exaggerates the differences that exist originally.

Professor Cronbach commented that a critical issue is the timing of assortative devices. It is inconceivable that the society is going to treat everyone alike, that it will go in the direction of preventing a bright child from reading more difficult books than the other children in his class after he is bored with the standard reader. Sooner or later there are going to be different treatments. The issue is this: Granted that there is a rolling readjustment of individual differences—and even though the stability is considerable after age 30 there is rolling readjustment in functioning all the way through the life span—should assortative mechanisms be employed at age 6? or should one insist that there be no differentiation before 11? or even that there be no differentiation until people have voluntarily dropped out of the system?

## FURTHER COMMENTS ON HERITABILITY OF IQ

In response to a question about the reliability of the statistical data on heritability, Professor Jensen indicated that he had compiled all the kinship correlations ever recorded by Burt in order to identify inconsistencies and errors and found twenty such discrepancies. For example, at times Burt reported a correlation without giving the sample size; in another article he would report the same correlation with a sample size indicated; and in still another article he would report the correlation with another sample size. This leaves open the question as to what the sample size really was and therefore what the standard error in the correlation is. To avoid this problem in the future, Professor Jensen recommended that at least for data that are hard to obtain, the journal that publishes the article should keep all the original data in its archives, including samples of the tests that were used. Statistical techniques for analyzing such data have been changing over the last twenty years, and many of the original data that Burt collected would now be quite valuable for reanalysis if they existed.

Professor Jencks suggested that a particular criticism that might be made about the Burt data concerns his method of obtaining IQ scores of the parents of the children for whom he had scores. In some cases there were school records for the parents, which was satisfactory, but in other cases their intelligence was assessed on the basis of interviews, which inspires little confidence in the accuracy of such measures. Dr. Eisenberg added that such a method is especially suspect if the interviewer knew the scores of the children of the parents being interviewed. Professor Jencks stated that Burt was conscious of that danger and tried to be careful in that regard although one cannot be sure that his efforts were adequate to prevent contamination.

Professor Griliches asked whether there is now enough additional data so that one could disregard the correlations compiled by Burt. Professor Jensen reported that there is only one correlation that Burt derived that has not been duplicated by someone else, the correlation between siblings born in cousin marriages. Genetic tradition would predict that these siblings would show a higher correlation than ordinary siblings and Burt's data confirmed this. This is an interesting correlation and hopefully it will be duplicated.

Professor Jencks felt that Burt's data are more impressive than others in several areas; if they were to be disregarded one would have little confidence in assessments based on the other studies by themselves. For example, the other data on siblings reared apart are not very good, and Burt has more data on identical twins reared apart than everyone else combined. Professor Layzer reinforced the point that there is no substitute for having the data and being able to examine them. In his view there are very serious internal inconsistencies in Burt's data which cast grave doubt on their validity.

Commenting that the data most useful from Burt concerned monozygous twins reared apart, Dr. Cavalli-Sforza pointed out that two-thirds of the cases in Shields' data on twins reared apart were reared with biological relatives, so the separation was very poor. This affects very deeply the correlations that can be computed. One question is: were Burt's data comparable to Shields' data, that is, were a considerable number of Burt's cases also reared in families of close relatives? That information is not available, and that is a serious problem. The Shields data (as well as the others) show a lower correlation coefficient than Burt's, and for the Shields data we know that the correlations are too high because the separation was poor. This would indicate a lower true heritability than is generally stated.

In that connection Professor Jensen observed that the IQs of Burt's twins are very normally distributed; they have a larger variance than those of Newman, Freeman, and Holzinger, and when the correction is made to adjust for that variance difference, the Newman, Freeman, Holzinger correlation goes up to .76, which is only a .01 difference from Burt's .77 for twins reared apart. Burt does get a higher correlation for what he called final assessments or adjusted scores, which consisted of an average of two or more tests given to the children; that produced a correlation of about .82. Professor Layzer asked about the skewness of the distribution. Professor Jensen did not think that it was very skewed; a comparison of the distribution of twin differences found in the three other studies—Shields, Newman-Freeman-Holzinger, and Nielsen—indicates that the distributions do not differ significantly in skewness.

## RACIAL DIFFERENCES

Professor Jensen reviewed the evidence for the belief that the test differences found among racial groups are not simply superficial or artifactual products of the testing situation or of the nature of tests. Regarding the situational context of testing, one of the claims made is that the race of the examiner has an effect on the test scores of children. However, a number of studies lead to the conclusion that the race of the examiner is a negligible variable, at least in group testing, when the tests are given according to the standard procedures recommended in the test manual.

Another question about testing relates to the motivation of children in the test situation. However, a considerable amount of evidence indicates that in a test situation there is little difference in the motivation of majority and minority children. One way of testing the motivation of children is to assign a task which is quite easy but which requires their attention and cooperation and compliance with instruction to a high degree. In such tests no significant ethnic group differences are found, except that in a very large sample small differences were found in favor of Oriental children, who seemed to be trying harder on tests that involve speed.

With regard to the question of culture bias in tests, both external and internal criteria have been examined. External criteria relate to the validity of the tests for predicting scholastic and job performance, and studies indicate the same validity for black as for white children in college students and job applicants. In one of Professor Jensen's studies, in which scholastic achievement was predicted from a battery of cognitive tests, no differences in predictive validities were found when race was added to the prediction equation. Professor Jensen emphasized that there is no evidence that indicates a trend that the predictive validities of the standard tests now in use yield different results for minority and majority children for the purposes for which they are most commonly used.

With internal criteria of bias one is dealing with a different set of phenomena: one is looking at the internal consistency of test variance. For instance, if one group of children is not trying so hard, either answering at random or just to please the tester, this should lower the inter-item correlations in a test. In fact the inter-item correlations of the tests that have been investigated—the Lorge-Thorndike, the Peabody Picture Vocabulary Test, Raven's Progressive Matrices—are the same for blacks and whites. Another important criterion of internal consistency is the rank order of item difficulties. Referring to the Peabody Picture Vocabulary Test, Professor Jensen indicated that when P values (percent correct on each item) are correlated over test items, the correlations between black and white groups are very high; the rank order of item difficulty for black and white children is about the same. That raises the question of whether the test is sensitive to differences of rank order, and indeed it is. When the same test was given to white children in London schools the rank order of item difficulties in some items was markedly different from that of California school children. In other words, there are marked cultural differences in the understanding of the meanings of these words, but no such differences are found between blacks and whites. In fact, there is more discrepancy between the sexes in rank order of item difficulty than there is between children of the same sex of different racial groups.

When asked whether it would be possible to construct a test that showed no black-white average differences, Professor Jensen replied that while possible it would not be likely to have many of the properties of a good psychometric test. Professor Kagan maintained that one should not rule out a priori the possibility of designing a psychometrically legitimate test that would have minimal racial differences. Professor Jensen doubted that such a test would measure whatever it is that is called *g*—the general intelligence factor. The American Institute of Research made a large scale attempt to devise a psychometrically valid instrument for Negro ghetto children; they sampled environments, determined what the children do, what they know, and found that the most common experience of black children in the area was television viewing. Therefore, they developed a test in which many of the items dealt with television knowledge, reasoning problems based on television types of experience, and so on. This test predicted scholastic achievement in the Negro group, and the expectation was that blacks would do better than whites on this test since it was sampled from their experience and interests; however, when it was given to a white group, they scored a standard deviation higher than the Negro sample.

Professor Jensen reported that with Raven's Progressive Matrices Test, where a blank space has to be filled in by one of several tiles to complete a pattern, the correlations between race in rank order of item difficulty are even higher than with the Peabody. An interesting aspect of this test is that significant differences were found between black and white children in the errors they made. There seems to be some developmental or age-related factor in the choice of distractors on this test; a ten-year-old generally makes a different mistake from the one made by the twelve-year-old, and black children twelve years old tend to make mistakes like those of white children ten years old.

Another means of determining cultural bias is to factor analyze these tests at the item level, looking at the *g* loadings of each item, and compare these loadings with the extent to which these items discriminate between blacks and whites. It turns out that the items with the highest *g* loadings discriminate the most—that is, whichever items have most in common with all the items in the test are the items that discriminate the most between the black and white groups. These items are those that require some kind of mental manipulation: one cannot arrive at the right answer simply by looking at it or by calling on something previously learned but rather must apply reasoning in the test situation. When asked how *g* is operationally determined, Professor Jensen replied that all the test items are intercorrelated, and the principal components—which are orthogonal factors or dimensions that account for a large proportion of the covariance among all the tests—are extracted from this matrix of intercorrelation. The first principal component, *g*, is the largest of these factors, the one that is most general to the battery. One can examine subparts of the test to see how much of this *g* a particular item has—that is, how much the item correlates with what is common to the whole battery.

Continuing with the issue of the developmental aspect of test differences, Professor Jensen suggested that the differences are not purely experiential or random but involve some kind of growth sequence that can be seen in a great variety of tests. Results in a steeply age-graded figure-copying test for four different ethnic groups—Orientals, whites, Mexicans, and blacks—show a full standard deviation from the top to the bottom group. The Oriental group, which is not the highest socioeconomic group, is at the top, two Negro groups from different communities and somewhat different social levels are at the bottom, and the white and the Mexican group (which is the lowest group on a socioeconomic index) are intermediate. The differences in terms of age or grade are quite large. The black fourth graders, for example, perform on this test on a par with Oriental first graders. The modal difficulties of each age with these tests seem to be the same in all of these groups, which one might not expect if cultural factors were determining whether a particular figure was difficult or not. Whatever factors are governing this, these kinds of difficulties seem to be the same in all groups except that they occur at different ages.

Commenting that children vary a great deal in artistic ability, a skill that does not correlate very much with other kinds of competence, Dr. Davis suggested that scores on this test might correlate as much with drawing ability as with reasoning ability. Professor Jensen indicated that this test is scored in a way in which the maturity of the drawing is separated out from its draftsmanlike or artistic qualities, and the two scores load on quite different factors. One child, for example, will draw a diamond with points on it in a very artistic way but in a form which is lacking in maturity. This test does correlate with reasoning ability because it has a high *g* loading in factor analyses with many other tests.

With regard to the suggestion that the distribution of mistakes in different racial groups is dependent on age, Professor Trivers asked at what age that no longer holds. Professor Jensen answered that these data have been compiled and analyzed only up to the eighth grade;

beyond that there is no information. Professor Herrnstein asked whether it is the case that if the groups are equated for percent correct then they will also have been equated for distribution of errors. Professor Jensen agreed; if two sets of individuals are matched for total number of errors on the test, one group black and the other white, one cannot tell these groups apart from pattern of errors or responses. Dr. Eisenberg asked what inference can be drawn from that fact. Professor Jensen suggested that whatever cultural factors went into the acquisition of the knowledge and skills tested by the test, they have the same relative probability of being acquired by the two groups; otherwise the rank order of these kinds of things would be mixed up. If anything of a cultural or environmental nature is depressing the scores of black children relative to white, it is depressing all of them uniformly.

In Professor Jensen's view it would not seem to be a cultural effect unless it is hypothesized that the culture depresses the learning of every kind of item that goes into these tests about equally, which is not likely. Referring to the British-American comparison, where the likelihood of exposure to different words is different in the two cultures, Dr. Eisenberg commented that if the black culture were such that the same words were used but less often, and therefore the likelihood of encountering them is reduced, one would get the same effect.

Regarding the question about verbal and nonverbal tests, Professor Jensen reported that these tests show about the same difference between whites and blacks. Although one sees the suggestion in the literature that blacks do better on verbal than on nonverbal tests, there is no clear evidence for that assumption. He indicated that he had tried to make a stringent test of this by matching a verbal and nonverbal test for item difficulty in a large white group, so that the items were perfectly matched for difficulty, percent passing, and so on; he then gave these two tests to a black group and the items were matched in the black group as well: there was no mean difference between the verbal and the nonverbal test. Mexicans, however, did significantly better on the nonverbal test, which might be expected in a bilingual group.

Dr. Morison asked about the general experience regarding Orientals—whether they are on the whole superior to whites. Professor Jensen indicated that in California they are superior on the nonverbal test and about the same on the verbal, except for immigrant children, who are much lower on the verbal but still superior on the nonverbal test. Dr. Davis asked if there is any information as to whether this is characteristic only of the selected Oriental population that has reached this country or of the whole population. Professor Jensen replied that the Wechsler test works very well in Japan and that the norms look very similar to those of the United States, with only a few minor changes in the items; this suggests that the Oriental population in this country might be a select group.

To illustrate the breadth of racial differences, Professor Jensen then discussed some studies of reaction time. From the time of Galton it was believed that reaction time—the term now used is "information processing capacity"—might have some connection with intelligence or alertness. Studies dating back to 1885 show a relationship between reaction time and the amount of information involved in the stimulus that has to be responded to, indicating that increase in reaction time is a linear function of the amount of information. Apart from the question of whether or not it is correlated with IQ, reaction time constitutes a specific behavioral performance which can be accurately measured.

In recent years investigators have employed simple tests of reaction time—for example, showing pairs of letters and asking people to indicate as quickly as possible whether they are the same or different. Professor Jensen described a test that he designed to show changes in reaction time as a function of information, one that was so simple that subjects would not have a subjective feeling of any increase in difficulty of the test. It consists of a push button which controls a light and at a warning signal the subject is told to turn off the light as quickly as possible by releasing the button. As additional light-button combinations are added, measurements show a linear increase in reaction time. When this test was given to 100 white and black youths who were matched on a psychometric test as well as on age and on education, no difference was measured at the lower level but there was an increasing difference as the amount of information increased. Since the test on which the two groups were matched was a rather culture-loaded test, if the cultural hypothesis were true one would expect that on a less culture-loaded test the blacks should do better than the whites. What happened was that they regressed to different means.

Professor Freeman suggested that an alternative explanation is that black children do not move as quickly and therefore they are doing less well on IQ tests because they need more time. Professor Jensen acknowledged that possibility but pointed out that many IQ tests are not timed in any case. In his view it is reasonable to conclude that this test reveals a racial difference: two groups matched for IQ show a significant difference in a behavioral performance that is very difficult to explain in terms of cultural factors. This raises the probability that IQ scores are not simply highly specific to cultural loadings in tests. Professor Jensen indicated that if one does a multiple regression of four items derived from reaction-time measures—the variance of the reaction time, the person's variance around his own mean, the reaction time, the slope, etc.—and uses these to predict race, the result is a multiple $R$ with race of .41. This is a substantial correlation, showing that two groups who are very much alike in many ways (age, education, psychometric test), but who differ racially, can be distinguished with a discriminate function of .41 by purely behavioral tests.

Professor Jensen concluded that in the light of such data the usual cultural explanations of racial differences were superficial and other kinds of hypotheses had to be entertained. With regard to the cumulative deficit hypothesis, which is a main pillar of environmentalist theories—the assumption that IQ in any disadvantaged group declines over the years as a result of cultural and environmental factors—Professor Jensen reported that when an adequate study of this was devised, which ruled out all artifacts that occur in cross-sectional studies and even in some longitudinal studies, no evidence of any cumulative deficit showed up. The groups did not change in the relative numbers of children at different rank orders over the years from kindergarten to high school; and although one cannot get good *g* measures below five years of age, it would be difficult to justify the hypothesis that the environment suddenly stopped acting at five years of age.

Professor Trivers referred to the phenomenon of compensatory growth in animals whereby a difference in body size produced by an environmental manipulation does not maintain itself naturally in the absence of further environmental deprivation. If there were such a thing as compensatory growth in intellectual function, and if environmental deprivation had operated up to age five and ceased, one would expect that there would be the closing of the gap, not the maintenance of the difference. In his view, showing that the gap maintains itself instead of increasing is not an argument to support the conclusion that it cannot be due to cumulative environmental effects. Professor Griliches maintained that this type of comparison is not applicable to the IQ issue because the IQ scale keeps changing; there is no constant metric, it is rescaled relative to different populations. Measures of IQ at age 9 and at age 5 are not like kilograms of weight because the scale and the standardization change. Dr. Eisenberg agreed that there are problems with using the body size model, but pointed out that it suggests one other possibility: that depending on the species and the time of environmental insult, there may be an irrecoverable deficit.

Dr. Davis commented, in relation to Professor Trivers' point, that if the difference in early cultural environment is the major factor responsible for the difference in IQ between blacks and whites as they enter school, then once they are both exposed to school some of that gap should close. Dr. Eisenberg stated that in that case one would have to assume that the schooling is equivalent and that the environments from that point are the same. Dr. Cavalli-Sforza added that such an argument is not valid if there are sensitive periods after which no compensation is possible. There is evidence that there are such sensitive periods for brain growth that are very precise.

Professor Herrnstein asked about data relating to blacks in Africa or other countries where they are not subject to discrimination. Professor Jensen reported that there has been a good deal of testing in Africa, and the reviews of that literature indicate that there are differences between whites and blacks; the scores obtained by blacks are a standard deviation or more lower than those obtained by whites. He added that there apparently are differences in different regions, but he was not aware of any comparative study of different geographical areas in Africa. Professor Lindzey observed that a considerable number of studies have been conducted in the West Indies. To his knowledge the only study that shows a reversal of the characteristic U.S. findings is an unpublished set of data collected in Antigua, which indicates a significant difference in favor of the black children in comparison with white.

The opposing view of racial differences—that the evidence is not nearly strong enough to justify the conclusion that such differences are partly genetic in origin—was represented by Dr. Cavalli-Sforza. Regarding the possibility that the IQ difference between blacks and whites is genetic, Dr. Cavalli-Sforza observed that Professor Jensen presented this conclusion in his 1969 article in the *Harvard Educational Review* as a hypothesis that should not be discarded without serious reason. However, in Professor Jensen's recent book, *Educability and Group Differences*, the hypothesis has become almost a point estimate. The statement is made that the genetic part of the 15-point IQ difference is between one-half and two-thirds. In Dr. Cavalli-Sforza's view the evidence presented does not justify this statement; rather than a conclusion it is only a guess. He suggested that Professor Jensen should give a clear indication of how he arrived at these figures so that the process on which the computation is based can be analyzed for its merits. (In response to this request Professor Jensen subsequently submitted a memorandum describing the basis of the quantitative aspect of this hypothesis—see Appendix II.)

Dr. Cavalli-Sforza indicated that, although a rigorous conclusion regarding the origin of racial differences is very difficult to arrive at, his guess would be in the other direction. He reported some recent evidence supporting this view which to him was very persuasive. As reported by Barbara Tizard in *Nature* ("IQ and Race," *Nature* 247: 316), a study was made in England of the IQ values of children of white, black, and mixed racial origin who were brought up in nurseries of high quality, with a generous staff-child ratio and plentiful supplies of toys and books—in other words, a "good environment." Scores on two sets of the tests showed little difference between the racial groups; in fact, the black children showed a slightly higher IQ mean than the white children, and when black and mixed-race children were put together and tested against white children, they were found to be superior.

Dr. Cavalli-Sforza emphasized that this is only one study; it deals with small numbers, and the data have been published in a very preliminary fashion, so one would have to reserve final judgment. But hopefully there will be more data of this kind forthcoming; until recently, this would not have been possible because there were not available for study sufficient numbers of black children raised in conditions that might be considered sufficiently favorable from the point of view of the analysis. Some confirmations that would be desirable, apart from having the experiment duplicated by independent researchers, would be to establish that these results can be obtained in America as well as in London, since it has been suggested that during the period of slavery black Americans have been selected against for intelligence, dull ones being preferred because they did more work. Dr. Cavalli-Sforza indicated that this kind of study is already being repeated in America and is giving essentially the same results.

Professor Griliches commented that if, as in this country, the white children who are given up for adoption are from a very different cut of the population on average, then without further analysis one really cannot draw any significant conclusion. Dr. Cavalli-Sforza pointed out that there was a test in this sample for distribution in social classes, and it showed a comparable distribution in those two-thirds of the cases in which the fathers were known. This, of course, does not solve the problem entirely, because it may be that blacks and whites have a different distribution in presently defined socioeconomic classes, and because different socioeconomic classes contribute foster-homes children disproportionately.

Dr. Cavalli-Sforza described a simple computation in connection with problems relating to the selection of the sample. Let us suppose that there are genetic differences for IQ among social classes such that the true (average) genotypic values of IQ in the lower and upper class are 90 and 110 in whites (and the same in blacks). Assume further that all white adoptees came from the lower class and that they have the average genotypic value of 90, while of black adoptees 3/4 came from the lower class and 1/4 from the upper class, so that their genotypic value is 3/4 (90) + 1/4 (110) = 95. Then there will be expected a five-point difference in IQ between blacks (95) and whites (90). This is about the observed difference of black and white foster home children in the Tizard data. Her values are actually closer to 105 and 100 respectively, instead of 95 and 90, but this phenotypic increase may well be due to the very good environments of the nurseries.

We are far from certain that genotypic differences in IQ for social classes exist, and if any, the one postulated above is likely to be an overestimate. One would obtain the same numerical result (an expectation of a five-point IQ increase in the black foster children over the white ones) by assuming that the genotypic value of the upper and lower classes are 95 and 105 and assuming that all whites came from the lower class and blacks came at random from the lower and upper class. These numerical illustrations are intended only for the purpose of meeting objections such as that of Professor Griliches and thus showing how biases of social origin may well be compatible with Tizard's observations and conclusions. This computation indicates that blacks and whites reared in essentially the same environment have the same IQ, and problems relating to the selection of the sample may account for only a few points. Thus the fifteen-point white-black IQ difference, if it is at all genetic, may be genetic only for a very small fraction.

Dr. Cavalli-Sforza then described a study of children of mothers who applied for prenatal care at two hospitals, one in Boston and one in Philadelphia. A comparison of the socioeconomic conditions of these two samples showed that conditions in Boston were much better than in Philadelphia, but in either city the whites and blacks were equalized fairly well for socioeconomic conditions. Test results at age 4 indicate that the difference between whites and blacks both in Boston and in Philadelphia is about five or six points, considerably less than the usual fifteen points, while the difference between whites in Boston and in Philadelphia is high, about twelve points. These data show that with socioeconomically comparable samples much smaller black-white differences are seen.

Professor Griliches felt that these data involve the same difficulty as adoption data. The people who go to the hospital for prenatal care may be very differentially selected from the two populations. He indicated that he had studied a relatively large sample of about 3,000 white and black children where a variety of regional and socioeconomic factors had been controlled, with a result of a ten-point black-white IQ difference; in his view, a very good socioeconomic standardization would probably eliminate much of that. But he felt that standardization in terms of those who apply to a hospital is not a reasonable selection and not likely to engender confidence in the data.

Professor Trivers added that in matching for socioeconomic class, one not only matches for environment, which is relevant to an environmentalist interpretation, but one also takes a subsample genetically if the populations are stratified by socioeconomic class; therefore, one has to know about the genetic stratification by socioeconomic class to know whether correcting for socioeconomic class is relevant. Professor Feldstein also objected to the inference that all one is controlling is environment when matching on the parents' SES. Dr. Cavalli-Sforza maintained that he did not cite these data as evidence but only as description of the differences between blacks and whites who were matched in some way by socioeconomic status. The data do not necessarily prove anything, except that when socioeconomic conditions are matched in this way the differences are not very striking.

Regarding the question as to whether the heritability within a group has any predictive value, Dr. Cavalli-Sforza referred to Professor Jensen's paper entitled "A Note on Within-Group and Between-Group Heritability," where at the beginning there is the statement: "A reasonable presumption (though certainly not proof) of genetic group differences seems to be related to the magnitude of the group difference and the heritability of the trait in question. . . ." At the end of the paper Professor Jensen states, ". . . at present, attempts to infer the magnitude of heritability between groups from the determinable heritability within groups is a blind alley." In Dr. Cavalli-Sforza's view, these two statements are irreconcilable. However, Professor Jensen maintained that the two statements are not contradictory. The reasoning in the first statement had been illustrated by the example of the mean difference between Watusis and Pygmies in height: when the phenotypic difference is so large and the heritability of the trait is high, there is a reasonable presumption that a genetic factor largely accounts for the difference. The other statement relates to the number of instances in which racial groups differ phenotypically and also genetically and the two differences are positively correlated.

Dr. Cavalli-Sforza went on to explain further why he believes that within-group heritability estimates presently given are too high. First he expressed increasing skepticism about Burt's data: they refer only to certain children in a small section of the population in London and are perhaps not representative. Second, the Shields data require a correction because of the great number of biological relatives among the adopting parents; when this is done, Dr. Cavalli-Sforza believes that heritability estimates will be considerably lower. He emphasized that even if heritability estimates within a group were 100 percent, the difference between groups would not necessarily be genetic. Moreover, there might be no phenotypic difference between groups which do differ genetically. Until a few years ago the heritability for PKU was about 100 percent, and now the environmental conditions have been manipulated in such a way that there is little difference between the phenotype of the PKU treated child and the normal child.

A major question relates to what is being measured by heritability estimates. Dr. Cavalli-Sforza distinguished between biological and cultural inheritance. He noted that genotype generally refers to DNA; a protein made by DNA can also be considered as genotype operationally if that protein is known qualitatively—that is, if its sequence of amino acids is known. Beyond that, everything that is observed is considered phenotype. The phenotype of learning ability and teaching ability will depend on DNA, and DNA is transmitted from parent to child. But the performance of the child will also depend on the effective teaching ability of the parent (which is phenotypic) as well as on his own effective learning ability (which is also phenotypic) and on the social and physical environment of his parents—his relatives, his sibs, his school, his teachers, his nutrition, etc.

Since the teaching ability of the parents will influence the performance of the child (although less so as children become more independent of parents and more dependent on the school), there is a phenotypic transmission from parent to child. There are other direct relationships between generations, because the social and physical environments are also directly transmitted from parent to child. This means that there are inheritances, mechanisms of transmission, that have nothing to do with DNA, although it is very difficult to separate, in such a complex pattern, the effect of transmission of DNA (genetic inheritance) from the rest ("cultural" inheritance). The existence of these alternative mechanisms of transmission is frequently ignored in usual heritability estimates; their effect is confounded with the genetic one.

The only way to estimate heritability properly is to separate the transmission due to DNA from the others by adoption studies. Therefore, analyses of MZ twins reared apart would be most useful, except that in a majority of cases the children are reared in families that are biologically related. Studies that do not use adoption necessarily involve both biological and cultural transmission in parallel, and cannot rigorously distinguish between the two. Cultural transmission alone can yield a set of correlations between parent and offspring, sib and sib, etc., that are very similar to those that would be obtained under biological inheritance alone. Therefore in the absence of adoption data estimates of heritability are largely unsatisfactory and might be entirely misleading. (Specific reference was made to the paper by him and Feldman, which appeared in *Am. J. Human Genetics* in 1973.)

Using the example of schizophrenia, Dr. Cavalli-Sforza called attention to the problems relating to much of the available data. Considerable attention has been given to concordance which in DZ twins is about 12 percent and in MZ twins about 40 percent. However, all this shows is that either there is a genetic determination of schizophrenia or the MZ twins have more similar environments than DZ twins, or a combination of the two. The only rigorous conclusion from this type of data is that there are environmental effects because MZ twins do not have 100 percent concordance. Only adoption data, which show that the children whose biological parents were schizophrenic tend to be schizophrenic, give rigorous proof for a genetic element in schizophrenia.

If additional data should confirm Tizard's study and establish that the black-white IQ difference of one standard deviation is not genetic except perhaps for an unimportant fraction, then, Dr. Cavalli-Sforza indicated, one would also have to conclude that another assertion of Jensen, that compensatory education cannot have an effect, is not likely to be true, and one should determine what constructive action might be taken. He agreed with Professor Jensen that the black-white deficit, whatever it is due to, appears before school age; possibly it would be useful to look at environmental differences that apparently occur in the same period and that might suggest mechanisms contributing to the effect. He pointed out two of these: the differences between single children and twins (five IQ points) or triplets (nine IQ points), and the IQ variance accounted for by birth order (Belmont and Marolla, *Science* 182:1096, 1973).

He offered the opinion that these differences might be explained if an important factor in the development of intelligence is the time spent with adults versus the time spent with age peers, and added that if it is true that the time spent with adults of the right quality is important, a policy inaugurated by Great Britain of starting compulsory schooling earlier (at three years of age) may be the kind of thing one should encourage. Professor Harrison agreed and commented that there is a good deal of evidence, obtained in Britain, for a large relationship between the amount of parent-child attention and socioeconomic class.

Professor Herrnstein pointed out that the deficit of individuals from one-child families represented an interesting deviation from this trend in the Belmont and Marolla study. Dr. Cavalli-Sforza agreed that his analysis did not account for the deficit of the only child and suggested that it is most likely the case that there are many factors determining these differences. Pointing out that it is at least intuitively plausible that most younger children and only children are born to older mothers, Professor Jencks suggested that this might be a causal factor for the deficit. Dr. Morison added that one reason for only children is some kind of obstetrical problem, which makes it impossible to have additional children, and in that case the child is likely to have suffered from anoxia, which would be a contributing factor.

In any event, Dr. Cavalli-Sforza maintained that group differences are essentially of no interest for educability. Unless the differences between groups were much sharper than they are, one would still have to test individuals in any case. In addition, as Thoday has suggested, emphasis on group differences, which may in fact not exist, might diminish interest in existing and important individual differences.

An important consideration is the relation of IQ to socioeconomic status. Dr. Cavalli-Sforza suggested that blacks are probably more worried about their being on average poorer than whites than about the suggestion that they are genetically inferior in terms of IQ. He referred to a family income graph, which showed that for some time the ratio of black and other races to whites has been going up, and that it stabilized about 1970. Dr. Cavalli-Sforza stressed the fact that scholastic curriculum is more important for income than IQ and family background, although they all correlate, and undergraduate enrollments of both whites and blacks have been going up, although blacks have always been lower. Because they have been going up, one might expect—and one does indeed find—that the ratio of average black to white family income as a function of age is highest for the lowest ages and decreases with increasing age. In Dr. Cavalli-Sforza's view this is in agreement with the idea that scholastic curriculum is important in determining income, and that increased schooling accounts for the increase in the incomes of blacks of a younger age; however, this also shows that it will be several decades before stabilization.

Professor Freeman suggested that there are some difficulties with some of the evidence cited regarding the socioeconomic position of blacks. He pointed out that family income data runs into the difficulty that increasingly large numbers of black families are female headed, so that the income of the man and the income of the woman, which in white families will typically be added together, are often separate in the black data. In fact the blacks have moved much more strikingly and have actually caught up to whites in many aspects of socioeconomic status. If one holds fixed the years of schooling, in the last three or four years the differences in income and occupational position between blacks and whites are very small, and in college graduates there are no differences at all. Most of the current income differences between blacks and whites are found among older black men, and there is some evidence that even their position has improved. In Professor Freeman's view, a detailed analysis of income and occupational data for blacks indicates that there has been a revolutionary change in their economic status in the past ten years.

Regarding the Tizard study, Professor Mayr called attention to the possible objection that the white children included in the study were of considerably lower intellectual status than the white population on the average; also, he questioned whether the blacks in this sample can be compared to American blacks because the study included a large component of West Indians, and a number of studies indicate that West Indian blacks score considerably higher in all sorts of tests than the particular West African blacks that entered the United States. Professor Jensen added that another problem with Tizard's study is the small numbers: there are N's as low as 7 in some of the crucial comparisons.

Professor Lindzey expressed the view that the study in general is very poor; there is so much lack of control and absence of detail that none of the data has any substance. Professor Harrison concurred with Professor Lindzey's assessment of the paper and expressed surprise that it had been accepted for publication by a scientifically respectable journal. Dr. Cavalli-Sforza maintained that the Tizard data are as good as most studies which have been published. He emphasized that in such a preliminary report one expects the data to be presented in a general way without detailed analysis. Even if the numbers were low, they are sufficient by current statistical criteria for the statements that were made.

Professor Herrnstein commented on the observation that in estimating heritability there are certain sources of bias—that, for example, in comparing MZ and DZ twins the commonly used formula assumes tacitly that the environmental similarity of the MZ twins is the same as the environmental similarity of DZ twins, and since that is a questionable assumption the resulting estimate of heritability is also questionable and probably too high. However, he pointed out that if the correlation between DZ twins who live together is subtracted from the value for MZ twins who are separate, then one has surely overcorrected for that source of bias, and yet using Shields' data with all of its inadequacies one still gets an estimate of heritability of the order of .45 to .5. In his view that would be a reasonable minimum value for the heritability; the estimate without any correction would be a maximum, with the true value falling within that range. Dr. Cavalli-Sforza again expressed dissatisfaction with the Shields study because it is based on data on MZ twins mostly reared with biological relatives. Professor Herrnstein felt that the environmental similarity, however great it may be, is bound to be less than the environmental similarity of DZ twins reared together.

Commenting that development is very complicated, Professor Kagan suggested that if the twins reared with biological relatives were firstborns their environments might be more similar than if one were in a dominant and the other in a subordinate position in the same home. He questioned the group's concentration on an independent variable and its acceptance that no matter what test is given it is probably measuring intelligence. Professor Jensen agreed that it is important to know the nature of the tests in question. For example, if the tests used in the Tizard study were tests of verbal maturity then there is little discrepancy with data from other studies.

On certain indices of verbal maturity—that is, not verbal reasoning but maturity of verbal development—it has been found that ghetto black children up to about age 3 or 4 are advanced over suburban white children. One such test is word association, in which there is a developmental sequence as to whether children respond with playing associations or function words, e.g., knife-cut or knife-fork. The highest level of that development is to give a word in the same form class—when given an adjective the child gives an adjective rather than a syntactic response. Dr. Eisenberg asked why that is not considered a reflection of intelligence. Professor Jensen replied that, like many of the tests given before four years of age, it does not correlate with later tests; in fact, in this case it correlates negatively with later measures that involve some kind of mental manipulation and reasoning.

Regarding the Boston-Philadelphia study, Professor Jensen underlined the criticisms already noted and questioned why one would put so much stock in a study with rather atypical self-selected samples than in other studies based on large random samples of school children of different social class levels. He referred to one set of data where children in Berkeley schools were classified into three socioeconomic groups and the results were consonant with several other large studies in which it was found that low SES white children still obtained higher scores than high SES blacks. In response, Dr. Cavalli-Sforza commented that the Berkeley population is probably even more self-selected than the Boston-Philadelphia one because there are so many children of academics there.

In response to Dr. Cavalli-Sforza's reference to his comments on compensatory education, Professor Jensen emphasized that he had never stated that compensatory education cannot have any effect; his statement was that the compensatory education programs at the time that the article was written had not had an effect—such was the conclusion of the U.S. Commission on Civil Rights, which had done a national survey on all of the major compensatory programs up to that time. Professor Jensen added that to his knowledge no survey has yet shown an appreciable effect for any compensatory program that could be replicated on a different group of subjects or in another school. Dr. Cavalli-Sforza commented that in his *Harvard Educational Review* paper Professor Jensen presents the "threshold hypothesis" to explain why compensatory education has failed; it can only keep failing if that hypothesis is true.

Professor Griliches called attention to the tremendous differences among whites across states. While one could develop a genetic hypothesis as to the settlement of the U.S. by different types, there are large systematic differences in IQ scores that do not seem easily explainable by a purely genetic model. One example in the U.S., in terms of motivation, success, ability to operate within the educational system, etc., are the Mormons, a relatively random sample of whites in nineteenth century America who went to Utah, created a culture out of whole cloth, and had an appreciable effect in two generations. Professor Jencks questioned whether it truly was a random sample. Professor Griliches replied that in that case one would have to ask what kind of selectivity was operating and could it have produced such a large effect.

Dr. Eisenberg commented that this question relates to Dr. Cavalli-Sforza's point that all one can say at present is that genes code for proteins and that from that statement to the complex of performance that one somehow assesses and calls IQ is an extraordinary leap, especially when one considers how the phenotype in much less complex traits is so dependent on relatively minor environmental variations. For example, there are *Drosophila* mutations that will not be observed if the fly is generated when the environmental temperature is kept to less than 25 degrees C.; the gene is there but there is no expression of it. In the same way there are likely to be many as yet unidentified factors that account for the enormous difference in performance among people. The difficulty with compensatory education is similar to that noted regarding the tests of abstract reasoning: they are both labels, not demonstrations. A weakness of the environmentalist position is that most of the statements made are exceedingly vague; until someone posits a hypothesis that is very specific and can be measured—that it is indeed time spent with the mother, or the way that time is spent, or how much talking the mother does—it will continue to be difficult to support the environmentalist position with concrete evidence.

Regarding the Mormons, Dr. Cavalli-Sforza indicated that if it is true that Mormons test higher than the rest of the white population, this gives some support to the theory that time spent with adults is important for IQ, because one of their beliefs is that women who are married should not work because it is very important that mothers spend all their time with their children. Professor Jencks pointed out the likelihood of significant differential reproductions, since polygamy was practiced among the Mormons for two generations, during which the rich had many children while the poor had none. Professor Mayr maintained that it was the especially capable people, not so much the rich, who produced most of the offspring, so there must have been a constant natural selection for a certain type of ability. Dr. Hoagland called attention to the pre-Columbian middle American tribes who established advanced civilizations in an early time. The Mayan group is still moderately intact, and he suggested that it may be worthwhile to study these people in connection with these questions.

In discussing racial differences, Dr. Davis suggested that one should keep in mind the modern biological definition of the concept of race—that is, any group within a species that has not been interbreeding for a long time with some other group, and that has therefore accumulated a substantial number of genetic differences. This delineation lacks a sharp border, unlike the definition of a unit in physics or chemistry; and as human beings have migrated faster and farther than most biological species, human races are becoming more and more mixed. What is called the Negro race in America today is quite different from the average in Africa; moreover, the average in Africa has no real meaning. In his book on *Race*, Baker has presented evidence that there is a wider range of physical types among the Africans than among the total European population.

If there are, in fact, accumulated genetic differences in behavioral capacities from one group to another, it may be of scientific interest to try to identify these differences, but most people would agree that in developing social policy one should concentrate not on group differences but on individuals. However, a large number of people in this country are now focusing on group identifications, assuming an *absence* of differences; and governmental policies are being established based on these identifications. In Dr. Davis' view, this approach has dangers, compared with emphasis on individuals, that are not receiving sufficient attention.

Professor Harrison addressed himself to the hypothesis that if there is genetic variation within a population there is likely to be genetic variation between populations. He agreed that as a general statement that is certainly true: most of the polymorphic systems, which by definition are variable within populations, do seem to show variations among populations as well. Nevertheless, there are two points to be considered. One is that some traits like stature (which probably have a genetic component) show geographical variation, but the actual level of variation on a micro scale is often much greater than it is on a macro scale. In other words, there is more variation in stature within Europe or within Africa than there is between Europe and Africa; so there does not have to be a broad geographical pattern associated with a system which is genetically varied.

The other point is that a number of physiological characters show little or no variance among populations once one has standardized for equivalent environments. Certainly there is enormous variation, for instance, in measures of heat tolerance in different populations, but once people are physiologically acclimatized to heat there is little variation. In contexts where human beings have great behavioral and physiological variability and plasticity, and where they can react to a particular environment by an appropriate phenotypic response, the selection pressures operating to produce racial differences are really very small.

Professor Lindzey called attention to an interesting paradox: with regard to heritability estimates of intelligence, many environmentalists would like to be able to prove that they are low—zero if possible—whereas geneticists would look upon the presence of a substantial additive variance component as providing one basis for arguing against the likelihood of genetic differences among races in modern society. If in fact there is a large component of additive variance, rather than dominance variance, that suggests that there has not been heavy selection pressure for intelligence.

Professor Jensen observed that the British geneticists Jinks and Eaves report about 15 percent dominance variance in their heritability estimates, and so they argue that whatever is measured by intelligence tests has been under some recent selection, but it has not been severe selection. Professor Herrnstein observed that in a large study of Japanese cousin marriages, IQ showed a large inbreeding depression; of all the traits examined only a few showed greater evidence of dominance. Dr. Cavalli-Sforza commented that the reported 20 percent dominance of IQ is based on the difference between the sib-sib correlation and the parent-child correlation, which can be heavily loaded with cultural factors. In any case, most of the genetic variance is claimed to be additive; this would make it unlikely that measured intelligence has been recently subject to very strong natural selection.

Professor Harrison commented that while one might argue that behavioral ability might be of general advantage over the whole species, and therefore be selected for identically or very nearly identically in many different places, it is important to keep in mind even for behavioral traits the possibility of a great deal of local adaptation, where selection may be operating against genes coming in because they are not locally advantageous. It is often argued that if an advantageous gene arose in one or two populations it would rapidly spread throughout the whole species; in fact, models of the flow of the gene under these circumstances show that it takes a very long time for it to spread through an appreciable number of the populations, even with strong selection. Therefore, on the basis of calculations of the effects of gene flow and the effect of selection, one could argue that populations separated over a long time would in fact differ significantly, even though there has been gene exchange between them and even though there is selection for the same gene.

Dr. Cavalli-Sforza commented that since the situation with respect to polygenic characters such as IQ is much more complicated, single gene calculations are not very convincing. Looking at this question from a different point of view, Professor Mayr pointed out that throughout human history there must have been a tremendous amount of constant natural selection in favor of at least maintaining, if not improving, a high level of ability; otherwise the human species would not be so different from the other primates. Dr. Cavalli-Sforza remarked in this connection that cultural evolution may have created an important threshold of selection: that is, society has evolved toward a state of high linguistic exchange, so that if people are not sufficiently intelligent to speak and understand with a minimum fluency they are automatically eliminated.

It may be that natural selection was very strong until that state was reached, perhaps 100,000 years ago, whereby those who could not speak or understand would not be part of the human species, but once that threshold was passed natural selection may have been less important. Dr. Davis felt that it is difficult to believe that man moved from hunting and gathering to farming, and from farming to living in cities, without substantial selection pressure for a different pattern of behavior. Dr. Cavalli-Sforza commented that there are many misconceptions about the urban transition. An interesting factor is that cities have never reproduced themselves until very recently; since urban populations keep dying out, this suggests that there cannot have been a great deal of selection in favor of qualities for urban life.

Dr. Eisenberg commented on the mental agility with which those on both sides of the IQ controversy explain away data that seem to be discordant with their position: the criterion for acceptance of data seems to be their matching one's a priori hypotheses. He reiterated his earlier suggestion that a useful activity for the group would be to try to define what kinds of data would be considered a decisive contribution to the resolution of some of these questions.

## SOCIOECONOMIC CLASS DIFFERENCES

A paper entitled "Notes on IQ Genotype and Social Class" (see Appendix III) was prepared for the seminar by Christopher Jencks and distributed in advance. This paper dealt primarily with the role of IQ in determining social success. Professor Jencks opened the discussion by describing his studies on the estimation of the heritability of other pertinent characters. He pointed out that there are certain practical difficulties in estimating heritability for IQ itself, although in his view they do not invalidate the effort entirely. Calculations derived from studies of MZ and DZ twins and siblings from Project Talent result in heritabilities of .67 for IQ, .45 for educational attainment, and .74 for career plans at age 23. But for the last the correlation between DZ twins and between siblings is so low that one has to postulate a negative term for the environmental component between families in order to reconcile the DZ correlation with a heritability of .73. A negative between-family environmental component is not reasonable.

The reason the number for career plans is out of line is that while the total variances are quite close for the MZ and DZ twins with respect to both IQ and education, with respect to career plans the total variance is smaller for MZ than it is for DZ twins. The implication is that with respect to the verbal formulation of career plans, some kind of mutual interaction between the MZ twins (more than between the DZ twins) is reducing the variance within families, and hence the total variance for the MZ twins. This suggests that the application of these types of techniques becomes progressively more hazardous as the trait being measured becomes more plausibly subject to interactions between the two people involved.

Professor Layzer questioned whether such a mathematical interpretation is at all meaningful, not only as applied to career plans but also to IQ measures. There are egregious inconsistencies and difficulties with the data on IQ as well: if one selects the data one can perhaps find a consistent set of heritability estimates, but in any case the theoretical basis for such an analysis simply is not there. Professor Jensen pointed out that the same anomaly found with the career plan heritability estimate is found with some personality traits that are even more obviously based upon some kind of interaction; dominance/submission is one of those that gives bizarre results when estimated in this way.

In any case, Professor Jencks continued, IQ is the only trait for which there are any reasonable data that will permit consideration of possible intervening mechanisms between genetic factors and socioeconomic success. As described in his paper, the figures on correlations between IQ and various forms of adult success vary dramatically, from a maximum of about .7 for educational attainment to something on the order of .3 or .4 for income. There is a difficulty with these data, in which one simply correlates the tested ability of adults with income or occupational status or educational attainment. Namely, there is a good deal of evidence that test performance in adulthood is to some extent influenced by the amount of schooling that one has had, and it is at least arguable that it might also be influenced by the kind of occupation one has engaged in.

Although there is no direct evidence that the latter hypothesis is valid, various studies support the prima facie notion that a long period of engagement in a particular occupation would have some effect on the kinds of skills that a standard IQ or adult ability test measure. In addition, the evidence suggests that school environments are relatively uniform compared to most of the environments that people encounter in the world; hence 6 to 16 would be the age at which the heritability of test performance would be maximum. After people leave school, particularly if they had been engaged in very different occupations for long periods of time, the range of environments and the size of their effects on IQ would be much larger. One would not expect a dramatic difference on this basis, at least judging by the effects of amount of schooling on test scores, but one might get a reduction of 10 or 15 percent.

This suggests that unless one has some direct evidence about the heritability of test scores in adulthood, if one wants to analyze the causal chain that leads from genotype to test score to adult position, one has to deal with the heritabilities of childhood tests and then look for relations to success in adulthood. Professor Herrnstein commented that since higher education and occupational status may very well influence IQ, and testing in childhood introduces the hazard of maturational variation, the sensible approach for such studies would be to use data from tests given at the age of 16 to 18.

With regard to genetic differences between the social classes, Professor Jencks referred to Herrnstein's famous syllogism and called attention to a formal path model which he developed to indicate that genotype influences IQ scores, IQ scores influence some form of occupational or economic success, and therefore genotype has some influence on SES. In his view, the logic of the argument in formal terms is almost irrefutable if one accepts the basic premises, which seem beyond serious debate with one conceivable exception—the assertion that the child's IQ influences his adult success in some direct way that will not be eliminated if one were to control all other conceivable factors. Clearly, such things as the status of the parents and the region of the country reduce the observed correlation somewhat.

However, in this regard one kind of test is quite instructive: one can look at the impact of IQ differences within the same family on educational attainment and occupational status. One can subtract the younger brother's values from those of the older brother for career plans and for IQ scores, and then regress one set of differences on the other to get what amounts to a within-family relationship between test scores and career plans. These relationships are always consistently smaller than those found in the general population, but they are never anything like zero. Hence it does not seem likely that by controlling family background characteristics one can make the relationship between IQ scores and adult success disappear entirely.

Another instructive exercise is to conduct the same kind of within-pair regression analysis within monozygotic and within dizygotic pairs. If one regresses the differences in educational attainment on differences in IQ scores within MZ pairs, which are by definition totally environmental in origin, one can in effect see whether a change in IQ score that is either environmental or genetic in origin has more or less or the same amount of effect on life chances. Professor Jencks indicated that at least with the Project Talent sample there is no difference; it is the phenotype that matters, not the genotype.

Assuming that Herrnstein's syllogism is theoretically correct, the question is what orders of magnitude one might attach to the numbers in the calculations. This problem can be approached with adopted children in two ways. One is to ask how large are the correlations between their IQ scores and the SES of the natural parent, assuming that they had been placed randomly. In a slightly more indirect way, one can determine if the correlation between the SES of the adopting parents and the IQs of their adopted children is of the same order of magnitude as that found in nonadopted populations. However, such studies have all been so different in character, small in size, and disparate in results that it is difficult to draw firm conclusions, and Professor Jencks found it possible to conclude only that the fraction of the correlation that is genetic is something between 25 and 75 percent, a rather unsatisfactorily large margin of error.

A question of somewhat more interest is to see how much of the inheritance of SES is likely to depend on the fact that the high status parents have genetically advantaged children with high IQs. These computations are summarized in Table 1, Appendix III. Professor Jencks observed that equality of opportunity as commonly defined means that opportunity should not be correlated with parental status except insofar as that is related to genotype: i.e., the world should be a meritocratic place in which people who have native ability can get ahead. The figures in Column 5 of Table 1 (Path with Genotype Controlled) give a kind of measure of inequality of opportunity (.32-.45 for educational attainment; .34-.48 for occupational status, and .30-.43 for annual income), and those in Column 2 (Respondent's IQ Genotype) give a crude estimate of the extent to which there is an IQ meritocracy in America today (.40-.55 for educational attainment; .28-.45 for occupational status; and .20-.32 for annual income).

With regard to the question of whether the system is becoming more or less meritocratic, Professor Jencks indicated that the available data show very little evidence of change in either direction. While it is always hazardous to assume that if situations have not changed they won't ever change, the evidence for a trend over time is very small. In Professor Jencks's view, the assumption of a genetic component in social class differences has little relevance in connection with the development of social policy if, as the evidence suggests, the environmental factors are incredibly difficult to change. If these factors were easily manipulated the implications would indeed be very different, but there is no experience that suggests that that is true.

Professor Jensen called attention to the statement in Professor Jencks's paper that heritabilities do not exceed the correlations between adult and juvenile scores. He suggested that such a conclusion probably has very little generality since, for example, the heritability of both juvenile and adult height is probably about .95, and the correlation between juvenile and adult height is no higher than it is for IQ. Professor Jencks maintained that if, as indicated in his paper, one assumes that the same genetic factors are operative in both periods of time, it is extremely difficult to construct such a model.

Professor Griliches commented that while he had no basic disagreement with Professor Jencks's analysis and conclusions, his own tendency is to work with large samples of data and not do much adjustment, in which case the relationships within any body of data look much weaker than they do after a variety of adjustments are made. Therefore, his correlations are all of about the same orders of magnitude as those of Jencks, but they tend to be perhaps ten points lower. He also pointed out that in the more complete models, almost all the observed effects of IQ operate via the schooling system, and the net effects are small. Moreover, the standard kinds of data do not distinguish between those who attended superior schools and those who attended schools with a low academic rating, though test scores have some relation to that factor.

The total explanation of observed income inequality is relatively small: only 50 percent of the observed variance in income can be accounted for in analyses of individual data. Also, the direct effects of test scores are rather small: a comparison of regression equations shows only a difference of .02 when test scores are included with other variables of age, color, and background. This suggests that one cannot predict the income of individuals with much greater accuracy if their test scores are known even when no schooling variable appears in the equation. Professor Griliches emphasized that the role of most of the presumed factors in socioeconomic success gets very confused except at clear threshold levels. Professor Layzer stressed Professor Griliches' emphasis on the limitations of what one can learn from such analysis; while it is very useful to have these data, it has to be borne in mind that the factors that are being analyzed have numerical but not quantitative significance.

Professor Jencks commented that the fact that the IQ mechanism in these models works almost entirely through schooling does have important social implications, which were referred to in his paper. It usually turns out that the correlation between parental SES and IQ is higher than the correlation between SES and performance—grades, performance on the job, etc. The result is a somewhat irrational system of selection, for which the school system appears to be the primary mechanism: it produces a higher correlation between parental status and achieved status than there would be if there were strict selection on job performance or on grades in school. If the use of tests was discontinued, and selection was based on grades, there would be a somewhat reduced effect of socioeconomic background on people's life chances. This consideration raises some questions about the virtues of a system of selection that weights IQ tests so heavily. Although the tests have enormous advantages, they do increase the transmission of status from parents to children in a way that is important to keep in mind.

In connection with the question of possible heterogeneity in the population units designated for study, Professor Harrison presented some data regarding the structure of a group of Oxfordshire villages. Records for 400 years have been analyzed to reconstruct the historical demography of these populations over that period of time. From the amount of exchange that has occurred among these villages and with the outside world, calculations have been made regarding the amount of genetic variety one might expect to find within the villages today. Also, one can derive some measure of the magnitude of the flow through the system through social mobility (whereby children have a different occupational status from their parents) and through marital mobility (whereby one chooses a spouse from a different social class). One can then determine the number of generations required to produce homogeneity. A deterministic model would predict that if the movement is random with respect to genotype, there should be no genetic heterogeneity among the different villages, or among the social classes.

Professor Harrison indicated that those predictions were tested by means of a house-to-house survey of the Oxfordshire villages by the use of a sociological questionnaire, IQ tests, anthropometric measurements, and polymorphic analysis. On the basis of polymorphic markers, these predictions seem to be fulfilled; there is no genetic heterogeneity in this respect among villages or among classes. However, the results for the metrics indicate that there is heterogeneity for IQ: there is an 8-point IQ difference between villages about three miles apart. This is due in part, but not entirely, to recent immigrants into the area. This heterogeneity is extremely significant, and it relates to the social class distributions of the populations: the village of Horton, which was significantly higher in IQ and stature, is likewise higher in terms of the proportion of people in Social Class I. A key question is whether the genetic constitution of one village produces a certain class structure that differs from that produced in another village with a different genetic constitution.

Professor Harrison reported that these data are being used to analyze within-family relationships, particularly in relation to social mobility and with respect to assortative marriage. Preliminary analysis indicates that in this population there is no assortative marriage for performance IQ; there is assortative marriage for verbal IQ, but it curiously exists only in two groups: those where both individuals were locally born, and those where neither individual was locally born. This suggests that even within relatively small populations there is so much heterogeneity that it is dangerous to generalize.

In conclusion, Professor Harrison indicated that since data were available on both the polymorphic systems and the metrics of this population, these data were examined to see if there were any association between the two. Comparisons of total IQ, performance IQ, and verbal IQ with ABO blood group showed that in one village there was a significant association, in which people of blood group 0 performed better than people of blood group A. It was discovered that such a phenomenon occurs in all except the two villages with the highest IQs. A complication arises when a comparison is made between the locally born and the nonlocally born: the phenomenon is present among the locally born in the males but not in the females, and vice versa among the nonlocally born. One possible explanation is that the correlations are solely artifactual. All other explanations necessitate the assumption of an association of one kind or another between the ability and the system: a pleiotropic manifestation of the same genetic locus, which seems inherently unlikely, or linkage disequilibrium, which is possible in such a small population but which would then have no predictive value for other populations. If that were the explanation, it would imply that a major gene responsible for IQ was located somewhere close to the ABO locus.

Professor Herrnstein described an interesting calculation relating to IQ and SES. Assuming that broad heritability of IQ is .8, and that the IQ-SES correlation is .5, it would be impossible to conclude that the IQ gap between social classes is entirely environmental. For, if the correlation between IQ and SES is .5, then the variance in IQ accounted for by SES is 25 percent; and if all of that is environmental, then one encounters the fact that there is only 20 percent environmental variance in IQ to begin with on the assumption of 80 percent heritability, and so one ends up with a logical absurdity. If one reduces the heritability figure to .7 and the correlation between SES and IQ to .4, the variance accounted for by social class on IQ is 16 percent and the environmental variance is 30 percent. There is no longer an absurdity. However, if 15 percent of the variance is due to social class separation, then there is only 15 percent environmental variance left over within classes—i.e., the differences between people within a class are highly genetic. One is then safe in estimating heritability from within-family data (sibling correlations, parent-child correlations) because members of a family share a social class, and there is apparently relatively little environmental variance within a class.

Professor Layzer objected to using heritability estimates in this fashion. In his view, because of the mathematical and biological structure of the problem, the appropriate interpretation of those numbers that are labeled heritability has not been and cannot be established. He commented that the idea that one can deduce facts about the real world from pure logic has not been accepted by the scientific community for several hundred years. Harvey Brooks maintained that to say that one cannot use logic at all to analyze empirical data is unreasonable. Professor Layzer stated that a valid argument has to be logical in structure, but logic by itself does not enable one to make the transition from one fact to another; for that one needs a theory. In the case of heritability estimates, the theory in question is, in Professor Layzer's view, defective.

Dr. Cavalli-Sforza expressed the view that in relation to social classes one should not use estimates of broad heritability at all. If there are differences among social classes in genotypic IQ it must be because there has been a selection for higher IQ, and what is relevant for such selections in the transition from one generation to another is the narrow heritability. Professor Herrnstein replied that there are two questions involved. For the transmission of SES from parents to children narrow heritability is indeed the relevant parameter. However, for the coefficient of genetic determination of SES—that is, what percent of the total variance is genetic—the broad heritability is relevant.

Professor Herrnstein observed that the studies reviewed by Professors Jencks and Griliches make the assumption that different parts of a range in one variable do not act differentially on the other variable. However, in his view that is not the case in the relationship between IQ and most of the sociological variables, such as education, income, eventual social status, etc.; some parts of the range of IQs, specifically the bottom part, are very much more predictive of outcome than other parts. The Project Talent data mainly leave out the bottom part of the range, which one needs for the full range. Professor Jencks agreed about the bottom but maintained that in the range of 85 to 130 the relationship is very close to linear. Professor Brooks noted some data on the scientific achievement of graduates of Cambridge University: for each discipline there was essentially a threshold IQ for admission, but above the threshold there was no correlation between IQ and subsequent scientific achievement.

Professor Herrnstein concluded by describing three kinds of data which he felt it would be important to have. The first relates to the social class separation of siblings as a function of IQ. Since the siblings in a family share a home background, a ten-point difference in IQ between brothers should represent a larger genetic separation than the same difference between a pair of people taken at random. If genetic difference contributes to social status, then an IQ separation between brothers of a given size should actually predict a larger difference in social status than that IQ separation completely taken at random. Professor Jencks stated that his studies show the difference, in fact, to be somewhat smaller for the brothers—that is, controlling family background and looking at the within-family regressions of SES on IQ, the coefficients go down, but not dramatically so.

Professor Herrnstein indicated that another interesting study would be to see whether children rise above or fall below their parental social class as a function of their IQ. Some data along these lines suggest that high IQ children rise in social class and low IQ children fall in social class. Another figure of interest would be the fraction of the total population of men in the labor force over time in the past 50 years. Professor Freeman indicated that something like 95 percent of white men from ages 25 to 64 are now in the labor force and that figure has been highly constant over time. Previously, blacks had higher participation in the labor force than whites, but especially in the last ten or fifteen years, the black labor force participation has dropped sharply, to about 88 percent. Professor Herrnstein suggested that if the meritocracy argument is valid—that is, that people will be crowded out of the labor market because the standards for jobs are becoming higher—it will first become obvious in that particular statistic.

## SOME PRELIMINARY CONCLUSIONS

Dr. Davis suggested several conclusions that might be drawn from the discussion of group differences. There is general agreement that from any determination of heritability within a group, however precise, one cannot rigorously conclude that differences of means between reproductively isolated groups are to any extent genetic in origin; one can only draw conclusions in terms of probabilities from various types of circumstantial evidence—and people vary a good deal in their estimates of the probabilities. This principle applies to racial differences, where there is no significant mobility between the groups and one can postulate cultural factors to explain the mean differences (though their effect, as Professor Jensen noted, would have to be very large). With SES classes, however, for which there is mobility from one class to another, it can be concluded quite rigorously that differences in distribution of IQ between classes have a genotypic component.

Professor Layzer commented that two uses of the word "probability" should be distinguished. One can use it in the sense of one's general state of mind, or in the sense of plausibility. Probability and plausibility certainly will differ from one person to another, and probability, in principle, should be estimable. In his view one cannot draw the inference that there is a probability greater than zero of genotypic differences between blacks and whites with respect to IQ; the probability of the existence of genotypic differences between blacks and whites relative to IQ is equally likely to be negative as positive. Dr. Davis maintained that using the word "equal" introduces a numerical consideration that is difficult to support. Using a medical analogy, he indicated that a physician approaching a patient with an illness would look at the data and decide that it is more likely to be one thing than another; he cannot assign a number to it but he bases his medical action on the probabilities.

Another important point, Dr. Davis continued, was the observation that whether such differences are environmental or genetic in origin is irrelevant under some circumstances but of considerable importance in others, depending upon the corrigibility of the environmental or genetic components of the differences that are presenting a problem. Professor Jencks had indicated that environmental intervention is as difficult as genetic intervention, but Dr. Davis suggested that while there are very difficult problems in connection with improving and manipulating the present environment, as a generalization the statement that the environment is harder to control than genetic factors may be too broad. For example, it seems intuitively likely that the correlation of IQ with occupation or income is considerably greater today than 100 years ago because of an environmental change by which higher education became much more widespread.

Professor Jencks expressed doubt that the order of magnitude of the correlation between occupational status and IQ 100 years ago would be very different from what is found today. However, whether that is true or not, there is no question that there may conceivably be a set of environmental differences between social classes whose effects would be relatively easy to eliminate by environmental intervention; and indeed, many have tried to construct a set of environmental manipulations that would neutralize the environmental effects of coming from a low status background. If their view of the character of the environmental differences had been correct, presumably the interventions would have worked better than they did. That is arguable too, because one could say that the interventions were ineptly managed; but in Professor Jencks's view the problem was that the perception of the environmental problem was faulty.

Dr. Davis suggested that if there is a large genotypic component to these differences, the range that one can hope to change by environmental manipulation is small, and this may be the main reason for the disappointing results of environmental interventions. Professor Jencks replied that if the simple intervention model were correct, when children from highly unequal backgrounds enter the much more uniform environment of school the differences should be decreased after a few years; but that does not seem to happen. One can construct a model in which the differences would remain—one could say that it was not additive, it might have been multiplicative, that the effects of family background continue to operate, and so on. Or, Dr. Davis suggested, one could conclude that 80 percent of this difference was genetic. Professor Jencks maintained that even if 80 percent were genetic, one would expect it to diminish; however, if anything it seems to increase.

Dr. Eisenberg emphasized that one cannot conclude that because an attribute is psychological or environmental it is readily manipulable, especially in this case where the details of the effective environment are so poorly understood. Dr. Davis indicated that he did not mean to imply that if it is environmental it is easily changed, only that, in principle, if it is environmental one can hope to make changes, while if it is genotypic one cannot, unless there are some radical new developments in genetics. Professor Layzer maintained that the most likely a priori case with phenotypically plastic traits is that it is both environmental and genetic in a complicated way. There are important genotypic differences in capacities, aptitudes, appetites, motivation, attention span, energy, etc., that are relevant, but at the same time these interact with a host of environmental variables in a complicated way, so that given these differences in genotype one can still produce very substantial differences in performance by changing the environment. Dr. Eisenberg added the qualification that emphasis must be nonuniform changes of environment.

Professor Brooks suggested that there is an implicit model that environmental influences are primarily additive; however, it may very well be that they are entirely or very close to serial in their effects, so that in fact an intervention later in the series may have very little influence because of the omission of an earlier prerequisite. Dr. Eisenberg agreed that there is some evidence for a serial effect.

Calling attention to a confusion between what can be done for an individual and what can be done over time for the dispersion of a whole social system, Professor Jencks emphasized that if the difficulty is environmental rather than genetic, there is no reason to expect intervention to be effective for an individual who is already at a certain age; but what one might expect for another generation of people is a different matter. Professor Griliches felt that one should not dismiss the possibility of successful environmental intervention. It is conceivable—and there is some evidence, such as the experience of Frankenstein in Israel—that there are subgroups of the population who have been in some sense screened out wrongly and who show a significant response to an environmental manipulation. Dr. Davis commented that the cognitive rehabilitation of the cases in the Frankenstein study slipped away after these students finished school unless there was continued further attention. This result suggests confusion between intense coaching during a given period and a general increase in level of intelligence.

Professor Jencks indicated that he remains pessimistic about environmental interventions that aim at raising test scores, but he is impressed by other kinds of interventions that have been more successful. For example, Project Upward Bound, which dealt with children with low test scores who came from low income families, while unsuccessful in terms of raising their test scores, did succeed in getting many more of these children into college than their older siblings, and they stayed in college at about the same rates as other people whose test scores were higher. A kind of motivational element appeared to have compensated for low test scores, and the net effect was that these individuals did receive college degrees and presumably had enhanced prospects for occupational status and income, even though their test scores were not appreciably increased.

Dr. Davis's last point related to another aspect of the question of environmental range. He suggested that some of the discordance that has characterized the discussion may be based on very different concepts of the potentially available range of variation. Some people hold that the genotype-environment interaction is of the same order of magnitude as possible genetic components, and so one cannot calculate the additive or even the broad heritability. On the other hand, Professor Jensen has cited evidence that the order of magnitude of interaction is relatively small. The question is, are there environments that will preferentially and substantially raise the IQ of a large fraction of the population that has previously been lower than another fraction of the population? The dietary treatment of phenylketonuria (PKU) is frequently cited as a basis for expecting a successful environmental intervention for any genotype that causes one person's intellectual capacity to be lower than some other person's; but in Dr. Davis's view PKU, with its monogenic effect on IQ, is a very special case.

Dr. Davis described three kinds of ranges of environment that are assumed in these discussions: (1) the actual range that enters into the observed distributions of people living in the world today; (2) the conceivable range that might be introduced if experiments reveal specific environmental alterations that will improve the overall performance of a whole species, or will selectively improve the performance of those who do badly in the present educational framework or in other parts of the system; and (3) a sort of theoretical, mathematical, infinite range of environment, in which one can find an environment that will specifically and extensively interact with any genotype (i.e., one can assume an interaction component of any size). Dr. Davis suggested that in the mathematical analysis of Professor Layzer, which claimed to eliminate rigorously the possibility of reaching a meaningful conclusion about heritability ("Heritability Analyses of IQ Scores: Science or Numerology?" *Science* 183: 1259-1266), it was tacitly assumed that one can assign any value to the range of interactions in the environment, rather than restrict it to something plausible; hence the analysis might not be too pertinent to the empirical question of human diversity.

With regard to the assumption that the genotype-environment interaction contribution to phenotypic differences is substantial, Professor Layzer emphasized that the point made in his paper is that if genotype-environmental covariance is present and appreciable, which is an assumption most people would grant for IQ, then it is mathematically impossible to perform the separation of the phenotypic variance into genotypic, environmental, and interaction components. The environmental effects, which Dr. Cavalli-Sforza has referred to as "cultural transmission," are so confounded with genotypic effects that it is impossible on the basis of conventional heritability theory to effect a separation. Only when adopted populations are used, and then only when effects like selective placement are absent or can somehow be corrected for, is it possible to make a valid estimate.

There is an operational criterion for when interaction effects are likely to be important. They are not important if the increment resulting from a given environmental increment does not depend strongly on genotype. If the environmental increment is a taped program of reading instruction, and the population is six-year-olds, interaction would not be a plausible hypothesis if the change in reading score after the administration of this treatment is essentially the same for all six-year-olds in the population considered. If the range of genotypes were small, one might expect that to happen; if the range of genotypes were large, one would not expect to find that result.

An analogous statement can be made about keeping the genotype constant and looking at a range of environments. Professor Griliches suggested that is part of the scale problem. Professor Layzer maintained that while that is a point that is often made, it is not valid: one can linearize only for single genotype, not for the whole range. One cannot do it simultaneously in both dimensions if one is looking for a range for both variables simultaneously. Professor Jensen added that there are some interactions that cannot be rescaled because they are disordinal interactions where there is a cross-over—not just a difference in magnitude of effect in the same direction but opposite effects.

With regard to the suggestion that PKU is an isolated example, Dr. Cavalli-Sforza pointed out that several other genetic defects have also been corrected in this manner. Dr. Davis indicated, however, that these are all monogenic diseases, while the issue of interest is the normal range of phenotypically observed IQs in a population, that is, the variance of an exceedingly polygenic trait. While PKU involves a single gene, producing a defined, correctable biochemical alteration, we do not know the biochemical effect of even one of the innumerable polymorphic genes that contribute to the normal range of IQ variance.

Dr. Cavalli-Sforza emphasized that even with a polygenic trait such as IQ there is already evidence that one can influence it environmentally; there is no basis for saying that because it is polygenic it cannot be influenced. Dr. Davis stated that no one would suggest that it cannot be influenced; the problem is the magnitude of the influence. Professor Jencks suggested that the critical question is not whether it is polygenic or monogenic but whether one can identify the relevant intervening variable. Professor Brooks asked whether there is an assumption that a polygenic trait will take the equivalent of a polygenic intervention. Dr. Davis thought that the problem could well be viewed in that way. Moreover, it might be useful to go further and consider the possible genetic mechanisms in terms of major theoretical classes.

Thus since all aspects of neural function depend on genes, some genes must influence intelligence through specific biochemical effects on the structure or on the environment of various kinds of synapses. The effects of such genes can conceivably be modulated by environmental intervention. But other genes must affect intelligence by directing the development of the specific wiring diagram of an individual brain. Some aspects of this development depend on stimuli after birth, and hence are subject to environmental influence; but others must be more rigidly determined. We have no information on the relative importance of these several classes of genes. But if those involved in rigid structural determination create a large fraction of the genotypic variance in intelligence, which does not seem unlikely, the implications for the search for environmental cures of low IQ would be discouraging.

Professor Jencks commented that one polygenic trait that affects IQ is skin color. The most common theory is that genes affect skin color, skin color affects the environment to which an individual is exposed, and thus affects IQ scores. If this theory is correct, genes are influencing IQ. But the mechanism for intervention is not obscure—namely, eliminate discrimination against people on the basis of skin color and the effect of that set of genes disappears. The number of genes or the complexity of the determinants of skin color is not a critical variable in determining how to intervene in that situation. Of course, if the genes that influence skin color had a *direct* effect on cognitive development, this sort of intervention would not work.

Professor Brooks noted that one form of environmental intervention, namely social discrimination, may be so highly specific and so large in its effect that it in fact acts a good deal like the dietary treatment of PKU. Dr. Cavalli-Sforza added that by transferring children from a poor into a good environment, one can raise IQ by almost a standard deviation. Dr. Davis suggested that the discussion has been confounding two problems: (1) how much the average score of disadvantaged groups (especially blacks) in this country can be raised by eliminating discrimination and thus removing an artificial handicap to the development of their capacities; and (2) if there is a mean genotypic difference, which would then be revealed, will the possibilities for specific gene-environment interactions allow us to discover environmental measures that will eliminate that statistical deficit.

Professor Jencks maintained that until the mechanism by which genes influence IQ is specified one cannot make any reasonable predictions about whether it will be easy or difficult to intervene. If one assumes a mechanism as complex as the polygenic model itself, it could be very difficult to intervene, but one could conceive of a situation in which there was a complex polygenic determination of some biochemical trait that was relatively simple to intervene in, or alternatively, a situation in which a single gene produced an effect that was extremely difficult to manipulate. Dr. Davis indicated that this relates to the question of whether our concern, especially in relation to present social policy, should be with the range of environments that are conceivable in the foreseeable future, or with the assumption that the genetics and medicine of the distant future will be able to find some way of influencing highly polygenic traits.

Professor Jencks emphasized that one should not overlook covariance. If there is no concrete notion about the mechanism by which genotypes influence IQ, one cannot rule out the possibility that those mechanisms are in fact subject to environmental interventions. One plausible model is that relatively small genetic differences lead people to select their own environments in various ways, and in the absence of a specific theory about how that kind of influence operates, about why it is that a given genotype leads to a given IQ at the other end of the chain, one cannot assess the difficulty or the ease with which one might intervene to break linkage.

Professor Jensen expressed the view that genotype-environment interactions would be more likely to be found with variables other than IQ. This may partly be a function of the way IQ tests have been devised; that is, they minimize the kinds of variance that would depend on interactions in order to get stability over the age range, to get a large *g* factor; thus all kinds of items that would interact average out, leaving only the lowest common denominator of ability. Therefore, Professor Jensen maintained that attention should be focused on trying to improve children's scholastic performance, rather than on trying to devise compensatory programs that are assessed in terms of how much they boost the IQ. Dr. Davis commented that two different possible social goals, in response to presently recognized diversity, should be distinguished: to maximize the self-fulfillment of each individual according to his specific genotype, which would reduce the variance some but perhaps not very much; or to do everything possible to minimize the variance.

Professor Layzer suggested that the socially important education issues relate not so much to variances as to thresholds: can the incidence of functional illiteracy be reduced through appropriate educational and social arrangements is the sort of question that is important, not whether it is possible to produce ten times as many Beethovens. Dr. Eisenberg indicated that the social equity issue applies primarily to the people who fall below the threshold for most occupations. Aside from whether it would be desirable, it is probably not possible to make everybody an atomic physicist, because that requires gifts that are not commonly distributed. The relevant issue is that numbers of people may be below threshold for many satisfying or income-generating occupations in society because of environmental disadvantage. With regard to the issue of maximizing opportunity for the fruition of special talent, he commented that Beethoven's genotype transplanted to an environment without the western musical tradition would hardly have yielded the same results; this emphasizes the necessity for a system of very differentiated environments, in which special talents are able to manifest themselves.

Dr. Davis asked whether the two aims of greater equity and greater cultivation of special talents are in any way incompatible with the belief that there is a large measure of heritability to the variance that is observed. Dr. Eisenberg replied that it is taken for granted that in some sense the range of abilities that people manifest, like other qualities of human beings, have a genotypic component; incompatibility comes about only if it is concluded that because there is a genetic influence it is a waste to invest effort at the low range of the scale. What is not valid is to conclude that the reason that blacks or those in lower social classes do poorly is so completely determined by hereditary factors that investment in change or alteration is not worth undertaking.

Professor Herrnstein replied that in his view it would not be a waste but one should recognize that there are difficult political questions, and these become even more difficult if the differences between social classes or races are to a substantial degree genetic. If it can be shown that the performance of the children of people in one class is lower than that of the children in another class largely because of genetic differences, and even if we would know how to compensate for this environmentally but at a great cost, that raises a question of politics. Society has to determine whether it is in its best interests to close that gap with the required resources rather than invest them in more gifted children, perhaps making the difference even larger. And even if the difference is environmental, if the required intervention is very costly the same question arises.

Professor Brooks noted that another strategy would be to look at the problem not as one of distribution but as one of entitlement. One could have a policy whereby as much of the resources as necessary would be invested to raise individual IQs up to a certain minimum; the remainder would be invested at the upper end of the scale, where the cost is relatively low because there are so few people, and the payoff is so high in terms of benefits.

Dr. Eisenberg commented that the social phenomenon is very complex and made more difficult because of the sociologic values attached to assignment. The problem of relating educational strategies to parental expectations and to social pressures, in a way that will provide a good educational environment for the children, is made more complex by the labels ascribed and the expectations produced. Present methods may not really permit us to see whether the introduction of an educational intervention produces a beneficial effect. Most studies show no evidence of benefit from special education for children labeled as "slow," but that is an empirical problem of present methods and consequences of labeling per se in addition to the limitations of whatever specific educational interventions are attempted.

Professor Griliches observed that since labels exist and have very large social currency, when one operates in an uncertain world without sufficient information, and yet has to make decisions using all available pieces of information, labels will not be overlooked as pieces of information. In allocating children into groups according to potential learning ability, if one acknowledges the inadequacies of test scores and subscribes to the assumption of some correlation of genotype with social class, he will in such assignments put some weight on social class. Professor Jensen indicated that he had done a study which showed that the amount of variance in scholastic performance that is predicted from a battery of tests of a wide range of abilities is not increased by adding information as to the child's race; race adds nothing to the information derived from these tests.

Dr. Eisenberg suggested that the complexity of the problem also includes the self-perception of the individuals involved. It may be illustrated by the history of the Harvard Law School, the earliest of the Harvard graduate schools to make an effort to increase black enrollment. In the beginning an effort was made to set up a special program to give these individuals extra coaching, and for a while this survived; later it was no longer acceptable because black students felt it to be patronizing. Consideration was also given to setting up a five-year medical program for disadvantaged black students and other ethnic minorities, which might have been sensible for people who had less of a background; but that idea was abandoned because it was unacceptable to these people. If, on the basis of the best test battery available, children assigned to one classroom were 80 percent black and to another classroom 10 percent, that would be perceived in ways different from the actual intent.

Professor Jensen commented that because of this reaction, certain kinds of intervention which are educationally beneficial, involving differentiation of classes, can be introduced in all-white suburban schools but are not acceptable in integrated school systems. Black children lose as a result. Dr. Eisenberg agreed and related another illustration: a study of the effect of drug treatment on behavior disorders in school children was to have taken place at two adjacent black elementary schools, but because it was perceived as behavior control it was not accepted. The study was therefore conducted at a parochial school, and already favored children received an additional benefit.

Given the long history of the way certain groups have been treated in this country, their unwillingness to accept what are intended to be helpful programs is, in Dr. Eisenberg's view, not unwarranted. He emphasized the importance of taking into account such social forces at any given point in history: the way people perceive what is being done is not to be disregarded. Dr. Davis agreed that at the level of political action, such self-perception is indeed an important factor, but he suggested that the appropriate role of the intellectual community is to try to analyze and clarify the objective realities of the situation. In this way we may be able to help accelerate the passage through a period when self-destructive and irrational actions are taken, and to decrease the situations in which black children are deprived of beneficial programs because of the political convictions of their parents.

Professor Trivers suggested that the intellectual analysis of the objective reality of racial differences is by no means complete, and he cautioned against disseminating any conclusions before the various issues involved are clarified. He stressed Dr. Davis's earlier point that in relation to the genetic component of the measurable differences in phenotypic characters, class and race are two very different phenomena. In the case of class differences there is the mechanism of differential migration, which will generate average genetic differences between classes. There is considerable evidence that in our competitive society there is differential migration; in advance, one would expect this to have a genetic bias in favor of people of higher IQ moving up. However, in order to explore the long term consequences of this differential movement, one needs to know about variance in reproductive success as a function of class, and this is not always taken into account in statements relating to the possible future consequences of differential migration.

In any case, that mechanism is lacking for racial differences, and there is no a priori reason for supposing that a specified racial group has more genes relative to high intelligence than another specified racial group. There is essentially no differential migration between blacks and whites in the sense of individuals moving in and out. Since blacks in this country are in fact a hybrid population, resulting from mating between the original Africans and Europeans, if their European forebears were nonrepresentative with respect to intelligence, there would be an effect on the population classified as black. However, there is no evidence that the mating might have been differential according to genotype of the whites. The only mechanisms that remain are natural selection acting differently on two populations of humans geographically separated for a long time, and natural selection acting differentially during the 300 years that they have been living in the same general environment as the whites.

From an evolutionary standpoint it is expected that natural selection in different environments will accrete different genes relative to a whole series of traits, and there is no reason to suppose that intelligence, whatever it consists of, is equally related to reproductive success in two different environments. One would therefore expect differences, but there can be no a priori presumption of white superiority; the mere expectation that natural selection favors different gene frequencies relative to any trait, including intelligence, does not allow the ordering of any two groups. Also, in a situation where two racial groups are brought together, and for over 300 years one is subjugated to the other and given an environment that would be expected to depress the phenotypic trait being measured—and with the subjugating group essentially inventing a test alleging to measure objectively the phenotypic trait—in Professor Trivers' view it is not a permissible rule of inference to decide that because the phenotypic difference runs in the same way as the environmental differences, a genetic component is therefore to be expected running in that same direction.

Professor Jensen has noted that in many studies a phenotypic difference found in geographically separated populations is presumed to have a genetic component. It must be remembered that these phenotypic differences are measured with the two populations geographically separate, and that assumption is based on a certain logic embedded in evolutionary theory. Thus even when the physiological responses to the two environments create a difference in a given direction, that does not exclude the presumption of a genetic difference in the same direction—if cold makes animals smaller, genetics may also contribute toward smallness in the animals in the colder environment. But for every such study, one can find others showing that when there is an environmentally induced difference between two populations of animals, especially in experimental circumstances, there is no presumption that the genetic difference runs in the same direction. The situation with blacks and whites is not comparable, and Professor Trivers maintained that one cannot infer which way the genetic component operates.

He expressed the view that the history of the 300 years in question, which may also have had some genetic consequences in terms of different kinds of selections, is not a trivial matter. It is not merely a matter of a lack of cultural amenities for the blacks. It includes over 100 years of slavery, with the destruction or at least the alteration of the basic human reproductive unit. It also includes a long subsequent period characterized by overt discriminations, some of a very brutal and presumably psychologically powerful kind, and often accompanied by manifestly unjust allocations of resources relevant to the development of traits like intelligence. In short, the environmental pattern is very broad and complex.

In closing, Professor Trivers commented on the assumption that natural selection is expected to favor different constellations of traits in different environments, not merely more or less of a particular trait. The situation is particularly difficult with a trait like intelligence, which is hard to define and for which there is a hierarchical scale. Professor Jensen has discussed a series of internal parameters of tests which demonstrate to him that such tests are not biased in any clear way against black children. However, Professor Trivers emphasized that on the basis of evolutionary theory, and even assuming a genetic difference, one has to take into consideration the possibility that a hierarchical test may not catch ways in which different groups are better or worse in given traits.

Professor Jensen expressed agreement with most of Professor Trivers' remarks but indicated that he did not feel that one should make strong assumptions about the effects of historical environment on IQ. In the analogy involving heat and cold and body size, the assumption is at least implicit that these environmental variations were directly and highly relevant to the trait in question. The effect of slavery on IQ is not known. While there is a tendency to assume that poor environments will automatically have adverse effects on traits that are valued, such as intelligence, this is not necessarily the case.

For example, the Chinese who came to this country a couple of generations ago had extremely poor environmental conditions and were discriminated against, but now the third generation is on a par with, if not above, the general population in academic aptitude. Also, American Indians score more or less intermediate between blacks and whites in scholastic tests despite very deplorable environmental conditions, which still exist. Professor Trivers agreed that there should not be any a priori presumption that particular kinds of environmental influences have enormous effects on a given trait; however, he asserted that if the history of intense discrimination and slavery had any effect on intelligence it would be difficult to imagine circumstances under which it was positive; in his view it was either neutral or negative. Professor Jensen suggested that the effect may have been negative through genetic means—for example, an unstratified society where everyone is held at the bottom has little basis for assortative mating for the traits that are enhanced by education and social stratification.

Professor Trivers asked whether there is any evidence for the notion that a history of discrimination has a positive effect on intelligence. Professor Jensen indicated that he did not believe there is, although some people suggest that the discrimination against Jews and Orientals was part of the reason that they are above the population average in intelligence and achievement, since it stimulated their drive to succeed and overcome difficulties. In Professor Jensen's view assortative mating was probably a much more important factor.

Calling attention to the role of cultural differences, Professor DeVore noted that while the conditions of most American Indian reservations are indeed deplorable, they are so in part because the Indian groups have chosen to remain in that environment despite government attempts to remove them; the basis is their fierce pride in their ethnic identity. Young blacks are handicapping themselves because of their effort to assert values different from those of the dominant culture around them: achievement orientations have a nonpositive value to them. The different self-images and motivations are very important in these considerations.

In connection with the influence of motivation, Professor DeVore referred to what anthropologists have called the nonlogical native mind. Groups in West Africa have been discovered who count in base 2, base 4, and base 10, depending on what they are counting, and perform quite complicated mathematical operations of a limited sort; however, they are unable to multiply. The reason is that they do not write; without a multiplication table that one can look at and memorize, this type of calculation is very difficult—in fact, there is no evidence that any culture has developed a way of multiplying in the abstract. In Professor DeVore's view this suggests that a lack of motivation in the home environment to begin the early literacy processes will show up as a systematic bias throughout—that is, that the so-called prelogical mind is simply the preliterate mind. It is quite possible that what is measured by intelligence tests is in fact an artifact of literacy.

With regard to the question of selection operating in the same direction as the environmental plasticity, Professor Harrison suggested several distinctions: whether the effect of the environment on the phenotype is adaptive or not, whether the response to an environment facilitates survival in that environment (e.g., acclimatization or immunization), or whether it is a response in which the phenotype is in some way reflecting a change of fitness. One would suspect that when the genetic mechanism goes in the same direction as the environmental influence the latter is an adaptive response; and that when selection operates in the opposite direction to the environmental influence the latter is a fitness response. Professor Trivers agreed that this distinction should be further analyzed, especially in the context of thinking about the extent to which a trait like intelligence might be expected to be insensitive or sensitive to environmental influences early or late, and why.

## CONCLUSION

Although time limitations precluded further consideration of the social implications of the biological and psychological information that had been presented, Dr. Davis expressed the hope that the preliminary discussion of these issues might suggest to members of the group, and to others, areas where further analysis and study might yield information that would be useful with respect to the development of effective social policy.

Professor Brooks, president of the American Academy, expressed the Academy's appreciation to Dr. Davis for the considerable effort he had expended in organizing and planning this activity. He emphasized that it is an extremely difficult task to conduct a dispassionate and analytical examination of such a controversial subject, and indicated that the systematic approach adopted in the seminars might well serve as a model for subsequent inquiries dealing with human diversity.

# Appendixes

# Appendix I: Measured Mental Abilities: Lingering Questions and Loose Ends

Lee J. Cronbach

**What trait is being measured by general mental tests? In what sense is "general mental ability" unified?**

There is a tendency to think of intelligence as a circumscribed, consolidated trait more or less comparable to height. But measured "intelligence"—general mental ability—is an index of performance. The performance of interest is a sort of average over a broad collection of tasks or demands called intellectual.

"General ability" is somewhat analogous to "efficiency." We do not expect to locate the efficiency of an automobile or a post office system in any of its units, nor by summing some measure on the units. What matters is that the elements work together to put out the right motions at the right time to accomplish a valued result. A stereotyped performance would be inefficient; adjustment of means to ends is required.

The February seminar was given Humphreys' definition of intelligence as the total repertoire of responses the person has at his command, and also Davis' proposed definition embodying assimilation, analysis and synthesis, and communication. There need be no dispute about these definitions: a test constructed to satisfy one

At the request of the seminar chairman, Professor Cronbach prepared this memorandum after the February meeting to call attention to issues relating to intelligence testing that might be discussed further at the May meeting. Comments by Zvi Griliches, Lloyd Humphreys, Arthur Jensen, and John Loehlin on a draft version are given in footnotes.

would correlate very highly with a test designed to match the other. Note the agreement to define ability as what the person is prepared to do at present, rather than as what he was born with or what he might do after some further treatment.

Binet's classic description of intelligence ought to be brought in: "the tendency to take and maintain a definite direction; the capacity to make adaptations for the purpose of attaining a desired end; and the power of auto-criticism." This reflects his research interest—long before he thought of making a test—in "attention and adaptation." It also is consistent with his later suggestions that if one had to sum up in one word what his tests measure, the word would be "judgment."

Binet is telling us how the intellectually efficient person *uses* his repertoire of responses. It is not a matter of simple information or simple conditioned response. The person sizes up the situation, accepts the goal set for him, and chooses an action that takes him toward that goal. He monitors himself, and corrects acts before they become errors. This involves a large element of "personality"; e.g., it takes a certain self-discipline to think before acting, and to find fault with oneself. Whether, in the face of difficulty, persistence or withdrawal is "more efficient" depends on the problem and, we would say now, the payoff matrix.

The tester wants to know what the person will do[a] on the average, over problematic situations, situations that sometimes call for persistence and sometimes for withdrawal, sometimes for mathematical precision and sometimes for good guesses. In order to allow the idiosyncratic requirements of various tasks to average out, he makes his test rather heterogeneous. Even provided that the single tasks are at least weakly intercorrelated, the total score will depend heavily on the common element and very little on any specific element that contributes to only one or two tasks.

How much one knows about, say, the geography of Canada does not enter the Binet definition. Yet the person cannot adapt or criticize without knowledge. The tester's solution has usually been to base his items on "culture common" content. That is, he has intended to ask the subject to reason about, or remember, or manipulate symbols and images that everyone of his age and community have been exposed to. This ideal is not always realized in practice, but it has been implicit in thinking about "general ability." The emphasis on adaptation has also meant a preference for items

[a]Humphreys' emphasis on a "repertoire" of responses is a reference to what the person could do if we elicited the best response. My stress on the actual performance is more operational and puts a bit more weight on personality ("might do" vs. "will do").

that include some obstacle or source of confusion. The efficient person should be able to function when the display is "noisy," perhaps as well as he does when the signal is clear.

If response to a test task is an orchestration of knowledge and judgment and temperament, it does not seem that one underlying physical element can account for superior performance. The complex, diverse performance we value in higher mammals is likely to be a resultant of dozens of physiological parameters, insofar as it is physiological. Neural structures, metabolism, sensory acuity, endocrines, etc., all play a role. But among healthy organisms of the same age, these parameters are not equally likely to account for the origin of *differences* in performances.

Despite the myriad elements to be considered, ability is in a sense unified. With minor exceptions, performances regarded as intellectual are intercorrelated. Over the population, abilities in arithmetic and language and music are correlated. Them as has, gets. Whether the effect lies in the genes, or in the encouragement given in the home, or in the confidence that the person who succeeded in one enterprise brings to the next, the person who is adaptive in one context has an above average probability of adapting in another context. The "cause" is probably a mix of attitude, energy, available interpretative schemata, etc. It would not be very rash to say that rote memory is the only "intellectual" function that does not track along with the other subdivisions of "intelligence." (The correlation of Mechanical Reasoning with Verbal Reasoning among high school boys is .55; among girls, .59.)[b] I shall return to the topic of subdivision of ability.

**In representative mental tests, what performances are called for?**

It has been suggested that the seminar would welcome more

[b]L.G.H. mentions data from Project Talent where school *means* were correlated. Such correlations tend to be larger than correlations for individuals. To quote L.G.H. ". . . the correlation between general information possessed by boys and mean reading comprehension is .915. The lowest correlation among the test means is .820 between creativity and mathematics. The creativity test was not an especially good selection. In contrast, most socioeconomic variables are correlated with these school test means less than .40. Starting salaries for teachers approach .50. In Table 93 there is a wider selection of test means reported. The lowest correlation in this set is .75 between reading comprehension 10th grade mean and mechanical reasoning 12th grade mean. Even mechanical information, however, is not much more highly correlated with mechanical reasoning, this being .787. Mechanical reasoning test stands out in this group as being somewhat different from the rest. Mechanical information means, however, are correlated .911 with arithmetic reasoning and .935 with reading comprehension. I interpret these correlations to mean that the selective factors that place children in different schools—and of course there are a whole host of these—are selecting primarily on the basis of general ability."

concrete examples of mental test procedures and tasks. The examples cited below will serve to emphasize some of the points regarding definition and bear particularly on the issue of cultural influence on test score. The best examples for our purposes are the Wechsler and the Lorge-Thorndike. The Wechsler has nearly the same structure from age 3 to age 60. While the Stanford-Binet is different in format, the net results are so similar that the SB author has advised using the Wechsler as if it were an alternative form of the SB. The Lorge-Thorndike is representative of the group tests used at school ages short of college admission.

The ten Wechsler subtests for school children are Information, Comprehension, Arithmetic, Similarities, Vocabulary; Block Design, Picture Completion, Picture Arrangement, Object Assembly, Digit Symbol. The first five are combined in a Verbal score, the last five in a Performance score. Wechsler's interest was clinical; he wished to understand the functioning of his patients along many dimensions. He therefore included Information, for example, as a specific index of acquaintance with the culture; he found it useful to identify cases where this was depressed relative to the reasoning subtests. A Wechsler interpreter is advised to consider educational and language influences in judging the person and also to recognize confusions of emotional origin. (Easier said than done.)

The Lorge-Thorndike similarly gives separate Verbal and Non-Verbal IQs. Modern tests such as this are minimally speeded; giving the children additional time does not alter standings appreciably. Special interest attaches to Figure Classification, Block Design, Matrices, and Embedded Figures. These and similar tasks are generally regarded, on the basis of their content and statistical properties, as representing the core of the general-ability concept. They require the person to break through a Gestalt, analyze out relevant elements, and recombine them to satisfy a criterion. They have no direct dependence on school lessons. These tasks epitomize the "Level II intelligence" of Jensen.

Jensen's Level I consists of tests of Rote Memory. The most common task for measuring this is Digit Span. The person hears a string of numbers ("2-7-5," "6-9-4-7"), at a rate of one per second, and then says back what he heard. Some people can handle longer strings than others. While Digit Span enters some versions of the Wechsler, for diagnostic purposes rote memory is given little weight in measures of general ability.

Over the years, concepts of general ability and the content of tests has shifted. Once it was thought that the tests were somehow getting at a "constant" parameter of the organism; if the ideal test could be found, it was thought, the child's standing relative to others would remain constant over the years. Now we speak of "performance" or "competence" or "developed abilities"; the data are on what the child is, here and now. An admissible hypothesis is that different underlying processes account for superior (inferior) scores at different ages, even on tasks that are similar in form. Thus, in Digit Span, superiority among adults may depend more on use of silent rehearsal and rhythmic grouping than on sheer attentiveness or neural retentiveness.

Tests came to be seen increasingly as predictive instruments. After 1930 the kinds of tasks that did not predict school success began to drop out of the group tests. Thus the Scholastic Aptitude Test, which assesses reading comprehension and reasoning about quantitative material, replaced the American Council Examination (which had been rather like a difficult Lorge-Thorndike test). The Verbal Lorge-Thorndike predicted grades better than the Non-Verbal and is often used by itself.

*In public controversies about tests, disputants have failed to recognize that virtually every bit of evidence obtained with IQs would be approximately duplicated if the same study were carried out with a comprehensive measure of achievement.* The seminar itself seemed not to come to attention when Humphreys reported that intraclass correlations of twins on "knowledge of the Bible" were indistinguishable from the MZ, DZ correlations for nonverbal reasoning.

**Broad achievement measures predict subsequent school success as well or better than general mental tests do. That being the case, what is the use of the mental test?**

One important function of the mental tests in the 1920s was to obtain a measure that did not depend on the particular instruction a person has had. The aptitude test for college entrance was contrasted with the subject matter examinations of the College Board, which required preparation in certain subjects, following certain syllabi. In lower grades, the mental test gave an appraisal of the student transferring into a school that the achievement test could not, since the transfer student had not followed the same curriculum as his new classmates. This does not seem so important nowadays, when achievement tests emphasize general reading skills, etc., rather than lesson content.

The mental tests play an important role in a clinical workup of, for example, a poor reader. If he is average in performance that does not require reading and vocabulary but much lower in reading, a treatment that zeroes in on particular subskills of reading is advisable. If both reading scores and mental test scores are low, reading skills may not be the first order of business.

At school ages, the proper function of the mental test is to help in choosing an educational treatment for children who are not likely to do well in conventional schooling. When a child is a poor achiever, and yet shows good performance on tasks that are unquestionably intellectual such as Matrices, the school ought to be challenged to find a new way to teach him. This was the motivation for Binet's work, and it is the common practice in the reading clinic. But the potential service in schools of the nonverbal measure has been lost to sight, since verbal tests predict better.

In horticulture, it is necessary to determine whether a plant responds better to an acid soil or alkaline soil: the plant's "capacity for growth" depends on the soil, i.e., there is a G × E interaction. In human development one can expect Phenotype × Environment interactions. The treatment right for one person may not be right for the next. The good Darwinian notion of the fitness of an organism to a certain environment was overshadowed by the idea that some people are generally more fit than others. The testing movement tried to single out the persons who will respond best to the mainstream experience, instead of trying to tease out interactions.

Richard Snow and I, working on this problem, have only now begun to get clear the appropriate experimental designs and analytic methods for detecting interactions. There have been enough past studies to indicate that interactions exist, but the empirical picture is confused. The most useful illustrative finding for this seminar is the 1940 experiment of G.L. Anderson. In his sample, fourth graders who had high IQs and poor pretest scores in arithmetic profited from meaningful teaching in the next grade. Those with good initial ability in arithmetic but comparatively low IQs did their best when drill was the principal teaching method. It is the difference between the achievement test and the general mental test that justifies interest in the concept of intelligence. The Anderson difference was a modest one and not uniform from class to class. At best, such differences and their implications will be difficult to tease out.

These interactions between school-age tests and instruction are, as I said, statements about the phenotype. We can say nothing about their possible genetic base, at least at present. The G × E term in human heritability studies has been small. This fact is irrelevant, since our concern is with *creating* interactions by experimentally standardizing environmental conditions. Data collected by distributing genotypes over two or three controlled, contrasting environments, each of which is uniform, will show stronger interactions than data from organisms reared in random mixtures of the same environmental components. It is from the latter that estimates of the G × E term have come.

This has implications for notions of "capacity." Following Darwin, fitness is fitness for a particular ecology. Capacity to form an antibody, to use Davis's analogy from the February meeting, is a conditional probability statement: given this particular insult, the organism has a certain probability of forming the antibody. With schizophrenia the pertinent environmental challenge is less specific, but the logic is the same: given the kind of stimulation commonly found in this subculture, this organism has a certain probability of an overt schizophrenic reaction. His capacity for schizophrenia would be greater or less in some other environment, presumably. If we could develop a brilliant etiological theory, we could perhaps decompose a probability of 0.20 derived from data on heterogeneous environments into a probability of .00 conditional on one environment, and of 1.00 conditional on another.

There is a counterpart qualification on the notion of "environment." Genotypes form a nominal scale, capable of being ordered only when we specify the environments in which they are expected to survive. Likewise, what is a "good" environment depends on the distribution of genotypes to which we refer. If there are P × E or G × E interactions, the ordering of environments must vary with the subpopulation under consideration.

#### Is it profitable to subdivide general ability?

Most kinds of mental tests intercorrelate to at least a modest extent; consequently, the total score on a mixed test is primarily a reflection of differences on various common elements (attitudinal or cognitive) running through all the tests. There is a tradition of defining the heart of the performance as a "general factor" *g*, now frequently spoken of as a "fluid" or "analytic" ability. Certain nonverbal reasoning tests are thought of as rather pure measures of *g*. But reasoning ability cannot be separated off from such attitudes as Binet's self-criticism, nor from the verbal-coding skill one uses in thinking through a "nonverbal" problem.

If we think of tests hierarchically, with *g* at the top, at the second level there are two useful categories: verbal-educational and spatial-mechanical. The v:ed tests (vocabulary, verbal analogies, reading comprehension, etc.) include the best predictors of schooling. One can work down to finer and finer subdivisions. Thus we can separate vocabulary from verbal reasoning, and below that can measure scientific vocabulary separate from artistic vocabulary, etc. A structure in this form does not tell the whole story, but tests on the same twig correlate more highly with each other than with tests on neighboring twigs, and the correlation drops again if we move over to another branch. The abilities at the higher levels can be thought of as a residue or by-product—e.g., not just the aggregation of words learned in many particular lessons, but also the whole frame of mind that makes it easy to amass a vocabulary.

Subdivided abilities have a role in guidance. A weakness in mathematics makes engineering a dubious career choice, for example. Subdivision is even more pertinent in diagnostic teaching, where very fine analysis of, for example, spelling errors leads to efficient reteaching. These kinds of subdivision are almost entirely out of touch with the kinds of subdivision that might be tied to genetic variation. Performance on Block Design consists of dozens of acts, some of them unobservable. These must have run through biological channels, but it is only speculation to suggest that some act is facilitated by more permeable synapses, or larger neural storage space, or the like. The maze learning of Tryon's superior rats was superior not because of some general biological quality but because in *his* maze the animal that worked well for food, used kinesthetic cues, and was not much inclined to explore was likely to succeed. In human behavior also, genetic differences may emerge more through effects on qualitatively different microprocesses, rather than by acting on some one parameter such as neural conductivity that enters every reaction.[c]

**In what sense are test items fair to persons from different cultures?**

There can be no quarrel with the report that the mean score on Test X is lower in Population 1 than in Population 2. This can be

---

[c]From J.L.: "The moral of the Tryon story, however, is that if you select for performance in a very specific situation, you will automatically select for *all* genetically based traits that facilitate performance in that situation—motivational, temperamental, cognitive, physical—hence it is vanishingly improbable that Tryon 'brights' will differ from Tryon 'dulls' on (say) just one parameter of brain function. But suppose Tryon had done it differently, had really tried to select for 'intelligence,' and had selected rats on the basis of their performance on a variety of mazes, jumping stands, problem boxes, barpressing tasks, etc., under a variety of conditions of motivation, environmental settings, and so on—in short, had selected rats for obtaining a high or low total score on the items of a rat 'general intelligence test.' The chances that the resulting strains might differ only in something like neural conductivity or brain mass would be greatly enhanced. It is this hypothetical experiment, rather than the one Tryon actually carried out, that we should use as a guide to thought, assuming that the situation which confronted man during his evolution was more like the 'rat general intelligence test' than like the highly specific Tryon maze."

specified in such operational terms as to be unarguable. The issues about comparisons of cultural or racial groups arise from interpretations. There was once a sense that mental tests measured an inborn quality. Young Jan Masaryk was once confined to a New Jersey institution for the retarded because as a Czech, speaking little English, he scored low on a mental test. His low test performance was a valid record; the inference was not. Abuses similar to this have been documented in recent times. Hypothetically, one can argue that differences among persons who have had "the same environment" indicate genetic differences; but environment has subtleties far beyond the reach of an SES index.

Does the lower test score of Population 2 truly mean that these persons are less ready to meet the demands of the school? Of the job? Such issues cannot be treated adequately in this memorandum. Let me say, summarily, that a prediction in the form "Persons with a score of x will fail in this college with probability y" is likely to be as valid in one social group as in another.

The evidence on job success leads us to think that tests for certain jobs often give information that, while relevant to success, probably *should not* be relevant. Suppose Population 2 scores below Population 1 on a reading test. The weakness in reading ability is a fact; it can scarcely be described as biased. If the training program a company runs requires much reading, the test is a valid predictor and the members of Population 2 do have a higher failure rate. But when we relate the same test to criteria collected at the workbench, among workers who finished the training, the reading test proves not to be a predictor. The bias, if any, is not in the measurement but in the design of a training program that placed a load on verbal ability; if less verbal forms of training are designed, poor readers will get through to the workbench and prove acceptable. This echoes the earlier remarks on interaction: are we sure that the persons who lack aptitude for the schooling we presently offer could not be educated by other tactics?

Thirty years ago Allison Davis raised the question of bias in mental tests. It was his view that most tests were loaded against the lower class child; a prime example was a vocabulary item requiring one to recognize "sonata." His group compared performance of middle to upper class whites in Rockford, Ill., with that of blue-collar whites. Some items generated relatively large differences, and some generated relatively small differences. Two main points emerge from the study, in my opinion. (1) The differences are remarkably consistent over different types of items; the lower group virtually never surpassed the upper group. (2) Insofar as there is disparity, the upper group has its greatest advantage on verbal items. Clearly, the blue-collar children would do better if instruction were less verbal. But this throws us back upon the value question, whether verbal abstractions do or do not constitute an essential route to man's control over his fate. The concept *freedom of the press* simply cannot be reduced to iconic or sensory displays for teaching.

The Coleman data show that "bias" in the mental test, if any, is not specific to that test. Differences are quite similar on achievement tests. (The superiority of nonverbal ability—relative to other scores—among northeast metropolitan blacks was an isolated finding not confirmed in other regions.)

Perhaps it goes without saying to this group. But let us write down: *no test is culture free*, because no action is culture free. This, one can argue, applies to something as unsocialized as the Babinsky reflex of the neonate, since in principle what the culture chooses to feed the mother during conception can affect the child.[d]

The poor performance on western tests of more remote people says nothing about their biological adequacy. It is not so long ago that civilization was at a peak in Alexandria, but today's typical Alexandrians would be behind western norms on a geometry test. The people of the Kalahari are not up to our norms on pencil-paper tests, but we are not up to their norms—by an amount that could spell disaster—on interpretation of environmental cues in the desert. Incidentally, someone said last time that musical talents are not dependent on the culture. There is at least one report of great superiority of Samoans on rhythmic discrimination, which the author attributed to the role of drums in the culture.

A nice question is suggested by findings of Gay, Cole, et al., in Liberia. If people do well enough when a problem is presented in a familiar context, and then do badly when a logically equivalent problem is presented in an abstract or foreign context, what do we conclude? One alternative is to say: see, they are entirely capable of reasoning; you just have to put the problem in a context they understand. The other alternative: reasoning in the absence of familiar cues or even in defiance of them is what constitutes adaptive ability.

---

[d]From J.L.: "Let us also write down: NO TEST IS COMPLETELY CULTURE BOUND, because every action reflects genotype."

From A.R.J.: "Of course, no tests are 'culture free.' I'm tired of this shibboleth that one sees in just about every psychology textbook, unaccompanied by the *real* question: how biased are culture-loaded tests? This is an empirically answerable question with respect to defined groups—races, social classes, sexes, national groups, etc. The answer is likely to be complex and multidimensional, of course."

**What are the possibilities of identifying new kinds of aptitude, distinct from general ability as traditionally defined?**

This kind of question can be given an affirmative answer only through successful development of some particular new kind of test that proves useful. One alternative is to capitalize on memory abilities, Jensen has suggested. Tests where one need only remember, without intellectual processing or adaptation, rather often show low correlations with general ability. And such abilities can be used to some extent in education, maybe as first stage techniques to equip children with concepts that can thereafter be used adaptively. But if teaching of arithmetic, for example, is by strictly rote methods, the pupil does not learn all that is acquired from meaningful instruction.[e]

A second candidate is "divergent thinking." Fluency of ideas (e.g., a test asking for several possible titles for a story) is only moderately related to success on "convergent" tests where there is one right answer. Once it was thought that creativity could be identified with divergent thinking, but there now appears to be a consensus that truly creative performance requires a nice balance of divergent resourcefulness and convergent self-discipline. If a person has good convergent ability paired with poor divergent ability, or vice versa, this quite possibly signals an emotional problem.

There is a long history of interest in "social intelligence." Some of the work has been cluttered with romantic ideas about intuitive *Einfühlung.* On the other hand, American operationism has generally produced rather artificial verbal tests of social judgment. There may be a future for tests of skills in teamwork. Some cultures emphasize achievement by the group rather than by the individual. One can devise tests requiring collective judgment, but research of this type has not yet addressed the question of group differences and their relation to individual differences and cultural differences.

Ability to learn may subdivide. There is some evidence that ability to learn from films, for example, is far from uniform among persons with similar general ability. Also, some people seem to learn better when closely directed and some do their best when given materials to work with in their own way. Any educational practice based on differential abilities has to consider whether it is best to capitalize on whatever ability is strongest or to try to build up the relatively weak ability.

There are philosophical questions about what is to be called talent.

[e]From L.G.H.: "[I see a gap] between the use of drill as an educational method and the functions measured by immediate memory tests. I simply do not believe that there is very much in common here."

One can place a greater or less value on rationalism, on the emotional aspect of thought; on productivity, on esthetic sensitivity; on individual excellence, on the welfare of the group; on innovation, or on conformity to traditional models. The mental testing movement, rooted in the Enlightenment, Utilitarianism, Darwinism, and Progressivism, has consistently opted for the rational, the productive, the individual, the innovative. It seems to me to beg the question to say that intelligence is not an index of human worth, that it should not be "disproportionately" rewarded.

What is valued, of course, is achievement rather than the buried talent, but this implies that whatever makes the achievement possible is valued. One can equalize a number of human rights, and conceivably can even equalize what each person is allowed to consume. But a society, even by distributing more educational resources to some than to others, cannot bring everyone "to the same level." Not unless an upper limit is somehow placed on what the fast learners are allowed to learn. Social change would have to be more radical even than that, to eliminate the value now placed on contributions that derive from learning and thinking. And any proposal to reduce the valuation placed on such contributions has to consider what such steps do to the viability of the society.

**How stable are individual differences in ability? What is their long term predictive value?**

Humphreys stated soundly that under most conditions of testing, variations in fatigue, incentive, demographic characteristics of the tester, etc., make little difference. Perhaps too great an impression of robustness was left. When a person is thrown into a new subculture he is not likely to be very efficient in meeting the requirements of that subculture. After he is at home in the subculture, his performance is more likely to be stable.

This is critical in the evaluation of early education. Children taken into a preschool program may show large pretest-to-posttest gains on any kind of task simply because they learn what it means to be set a task, and learn to cooperate with adults. Not only Jensen's anecdotal report, but reports by Palmer, Zigler, and McHugh show that familiarization, or modification of the testing procedure to minimize the impact of failure, improves IQs of young children. The average increase is 6-10 points, but the increase for some individuals or subgroups may be dramatic. My guess is that the same thing would be found in testing adults at Ellis Island. But I have to accept Humphreys' view that the familiarization issue is not very pertinent to the black-white mean difference through the school grades.

Humphreys mentioned the J.E. Anderson study of 1939, which argued that gains in intellectual growth were random and unpredictable. This is part of a recurrent argument that mental tests do not measure "learning ability." It is necessary to discount this argument on several grounds. While Anderson was able to account for test correlations by the random increment assumption, the data also fit models that make a contradictory assumption. Analyses we have made show that changes are not random. The regression equation relating age-18 score to true age-13 score can be augmented; true age-9 score and parent education, properly weighted, improve the prediction. There are serious artifacts in the attempt to predict "changes," to the point where R.L. Thorndike dismissed the entire line of inquiry started by Anderson:

> Perhaps it must be concluded that, unless and until scales are developed that are truly homogeneous in the functions that they tap at all levels and are truly expressed in equal units,[f] we will have to forego serious attempts to give a quantitative answer to the simple but tantalizing question: To what extent will the children who have grown rapidly in intellect up to the present moment continue to grow rapidly in the future?

Part of an individual's change over time reflects simple error of measurement. Some may reflect differences in some biological maturation at an intermediate age. Some reflect cumulative influences of the home or school. The figures L.G.H. handed out in February showing that scores at age 8 correlate only 0.44 with scores at age 18 struck me as conservative. He agrees that correcting for error of measurement would raise the figures, to 0.55 in this case. This sample may be on the low side, and the value of about 0.88 I have for another sample (small) is on the high side. For a third sample that L.G.H. supplies, $r$ is about 0.77. O.K., there is both stability and change.

One must be cautious in reading correlations of 0.80 or so as "high." *Suppose* that the correlation between IQ at 10 and IQ at 18 is 0.80 and that the s.d. of true IQ is 15. Then the s.d. of true IQ at 18, among persons who have the *same* true IQ at 10, is $\sqrt{1-(.80)^2} \times 15 = .6 \times 15 = 9$. I.e., if we think of the IQ range as 60-140, the ultimate range of persons who at age 10 are precisely at IQ 100 is from 76 to 124. A coefficient of heritability of 0.80 may likewise seem more deterministic than it is.

While the mental test was introduced with the idea of predicting

[f] L.G.H. thinks that the issue of units raised by Thorndike is not highly important. Z.G. considers it worth further emphasis.

success, in school populations it predicts no better than achievement tests. There is faint, very faint patterning; the verbal tests being better predictors of reading than the less verbal tests. L.G.H. has noted that "aural comprehension" tests are particularly good predictors. All the predictive correlations are fairly large.

**If we were to get dependable information on the size of genetic contributions to mental test scores, how would that influence judgments regarding social policy?**

There seemed to be agreement at the seminar that the precise value of the coefficients of heritability is not important. It was also agreed that an index of the size of genetic components is descriptive only of a particular population having a particular joint distribution of genotypes and environments. That is to say, it is a historical index on the situation as it was in a certain setting, group, and time.

There was little emphasis on the following corollary: if genetic contributions were found to be enormous, this would not rule a policy of environmental manipulation out of consideration. What was true in the past need not be true if new environments are invented or if the conditional E G distribution is altered. Presumably the heritability coefficient for PKU mental defect was very close to 1.00 until the preventive diet was invented, after which much of the variance was shifted over to G × E interaction (since not everyone who needs the diet gets it). It might be said that a high heritability coefficient should make us skeptical when someone claims that a certain intervention improves general ability.[g] But even if H were .00, there ought to be tough-minded evaluation of every social intervention. Except where theory is vastly better than it is in this field, one cannot judge an intervention in advance of tryout.

The genetics of ability has obvious scientific importance. Anything that genetics shows about biological mechanisms underlying ability or the details of G × E interactions could have great practical significance. I am inclined to think that such questions can best be pursued by studies on subhuman animals. (But J.L. rightly adds the qualification that some abilities can only be studied in man.)

Mayr suggested in February that conclusions on human heritability have large implications for policies related to social policies and programs related to reproduction, and I agree. So one question is whether the seminar considers policies intended to affect reproduction differentially to be within its charge. My guess is that eugenic

[g]J.L.: "One thing a high heritability coefficient might suggest is that one ought to try to invent new environments, rather than trying to decrease the variance on existing environmental dimensions. This seems to me a *very* important policy implication."

proposals will get no hearing in this generation.[h] More directly on the agenda of the seminar are policies related to educational practices and employment opportunities. I am unable to perceive any inference that could be made about policies in these areas from even the clearest evidence on heritability.

Policies have to do with allocation of opportunity: to receive tutoring, to enter college, to be hired as a statistician, to be retained on the job. By and large, in those jobs where poor performance can be recognized there will be strong pressure to remove the inadequate performer from the job. Policy questions about testing reduce, in large measure, to decisions about the advisability of eliminating persons from the path to a prospective role *early*, on the basis of predictions of failure. Shall we tell a person not to spend his life as a brain surgeon when he enters school? In high school? At the end of college? Or after he has demonstrated his lack of proficiency on his first five patients? The probability of a wrong decision goes down, and the cost (to the person and society) goes up, when we delay longer. When should the individual and social investment be cut off?

The second policy question has to do with unequal investments in individuals. This seminar can scarcely concern itself with the political question whether it is sometimes wise to invest tax dollars in educational programs that make no difference. Interesting policy questions do arise, once compensatory programs have been devised and shown to work. (There may be some who argue that no compensatory program could possibly work, but I find it hard to see what evidence could support that view.) If an effective form of compensatory education proves to be expensive, how much should be spent, and on whom? We will want to make an expensive kind of compensation experience available to those we predict it will help most. Is it not likely that the phenotype will predict better than the genotype? Is it not likely, then, that resource allocation will continue to depend on phenotypes? Loehlin presses me on this: "no reason at all why it should depend *solely* on phenotypes, particularly if we develop better means . . . of assessing genotypes."

In principle, college aptitude at 18 could be predicted by weighing together the child's score at 6 and a measure on his parents (as a proxy for genotype), better than from data on the child alone. As we come to know more and more about genotypes we could use them predictively, but see the sticky questions. Corabelle wants to be a

[h]J.L.: "I'll predict that *within-family* eugenics, the choice by parents among the possible genotypes they can produce, will have immense popularity within a generation of the time it becomes technically feasible to diagnose IQ *in utero* (or good looks or straight teeth)."

brain surgeon. You discover that with her genes the probability is 0.99 that she'll develop a fatal tumor at age 26. Is it sound policy to allocate resources with that in mind? Suppose you could diagnose with probability 0.99 that despite his present IQ of 100, 6-year-old Wellington will burst into a display of genius at age 16. Should we prepare him for that happy day with an extra ration of resources now?

From Z.G.:

> The basic question is not just the efficiency of a given system but also its justice. Assuming that there are "victims" of nature, how do we compensate them? Should a cerebral palsy youngster be provided with (a) additional training (does it work?), (b) subsistence for life, or (c) a visible high class job so that other CP victims would be induced to undertake the training themselves? We might just as well start discussing the deFunis case.

These last notes scarcely begin to frame the questions on which policy must rest. But they do express my unrest regarding an attempt to derive policy implications from a circumscribed set of facts. I would be much more comfortable to start from the other end, defining policy issues we propose to review, and deciding in the seminar what facts bear on each issue. I think we would find that "intelligence," as distinct from competence, has no special role in the policy considerations, and that the historical index of genetic variance for either intelligence or competence has no relevance, so long as we are prepared to alter the distribution of E in the society.

From J.L.:

> Obviously, the "historical index of genetic variance" doesn't strongly determine the future, if we are willing to redistribute E, but to say it has *no* relevance goes too far. If in some existing subpopulation the heritability of trait W is high, despite wide variation in environmental variables X, Y, and Z and a broad distribution of genotypes across them, this is a very relevant fact to policy makers who might contemplate redistributing X, Y, and Z in an effort to change W. No amounts of information about genotypes stops us from changing phenotypes by changing environment, or tells us what distribution of phenotypes we ought to have, but such information could help us allocate limited resources of E so as to maximize our rate of approach to our goals, or could direct and stimulate the invention of new and more powerful Es (as knowledge of PKU led to low phenylalanine diets).

# Appendix II: Addendum to Human Diversity Discussion

Arthur R. Jensen

In his discussion, Professor L.L. Cavalli-Sforza quoted from the following passage in the last chapter of my book *Educability and Group Differences* (1973), and offered the opinion that the estimate made therein was just a "guestimate" which was unsupported by evidence presented in the book's preceding nineteen chapters.

> In view of all the most relevant evidence which I have examined, the most tenable hypothesis, in my judgment, is that genetic, as well as environmental, differences are involved in the average disparity between American Negroes and whites in intelligence and educability, as here defined. All the major facts would seem to be comprehended quite well by the hypothesis that something between one-half and three-fourths of the average IQ difference between American Negroes and whites is attributable to genetic factors, and the remainder to environmental factors and their interaction with the genetic differences (p. 363).

This hypothesis was based on many lines of evidence reviewed in *Educability and Group Differences.* Its specifically quantitative aspect (i.e., that "between one-half and three-fourths of the average IQ difference between American Negroes and whites is attributable to genetic factors") was based on the following considerations.

1. If the heritability $h^2$ of IQ is 0.7 to 0.8 in the white population (which is the best estimate we have from consideration of the total

evidence), then the IQ variance attributable to environmental factors is $1 - h^2$ or 0.2 to 0.3. The standard deviation of the total environmental component of IQ thus would be about 6 to 8 IQ points.

2. If one assumes similar heritability of IQ in the black population, the standard deviation of the environmental component of IQ is about the same as in the white (i.e., item 1 above). (The evidence for IQ heritability in black populations, though not strong, does not suggest that it differs appreciably from the estimate in white populations. The existing correlations for twins and for siblings are highly similar for blacks and whites. This does not prove that the heritability is the same in both groups, but it makes it the most likely hypothesis.)
3. If white and black populations differ on average by some 15 to 20 IQ units, as the preponderance of the evidence indicates, then, given points 1 and 2 above, if it is hypothesized that all of this difference is environmental, it must be concluded that the groups differ by about two to three standard deviations in all the nongenetic sources of variance that make for IQ differences *within the groups.* But few would claim that the microenvironmental factors that constitute the *within-families* variance (e.g., birth order) should be included among the causes of the average difference between populations. It is the *between-families* variance, i.e., environmental factors affecting all members of a family, that constitute the kinds of environmental factors that are claimed to make for social class and racial group differences in IQ. The between-families environmental variance is about one-half to two-thirds of the total environmental variance within racial groups. This means that blacks and whites, on average, must differ by some 3 to 4 standard deviations in environmental influences if the standard 15 to 20 points IQ difference is to be explained entirely in terms of the kinds of environmental differences that account for IQ variance between families within each of the racial groups.
4. A variety of socioeconomic indices, singly and in combination, indicate that the average white-black difference is about one standard deviation or less—far from the 3 or 4 standard deviations of environmental difference that must be postulated by a strictly environmental hypothesis of the white-black IQ difference. In fact, on existing indices of the environmental and social class variables most often claimed to account for white-black IQ differences, the average white-black difference is about the same as the average absolute difference in environments among whites, i.e.,

about one standard deviation. In terms of these measurable (and potentially manipulable) kinds of environmental factors, direct measurements of the home environments of adopted children suggest that moving one standard deviation up or down on the environmental scale pushes the child's IQ up or down some 4 or 5 points.[1] If we consider these kinds of environmental factors which contribute to IQ differences within racial groups, we can account for only about one-third to one-half of the white-black IQ difference.[2]

5. Therefore, environmentalists now hypothesize the existence of as yet unidentified and unmeasured factors to account for the racial IQ difference, factors which must be claimed to make for IQ differences *between* racial groups but not contribute appreciably to IQ variance *within* groups. No one has clearly specified the nature of these factors, and in my book I simply label them "factor X." Factor X is, at this point, purely *ad hoc*, invoked to explain the IQ gap still left when known, measurable environmental differences are taken into account. Notions such as "racial alienation," "white racism," consciousness of being a minority, identification with a historically mistreated minority, etc. are attempts to characterize factor X. While these factors may exist, it has not been shown, independently of the particular racial difference which they are invoked to explain, that they have any effect on IQ. And one may wonder why they do not apply to other minorities, such as Jews and Orientals, who also have been subjected to discrimination, etc., but who score at the national average or above on standard tests.

   *Educability and Group Differences* reviews all the claims of environmentalists about each of numerous factors said to explain the lower average black IQ—inequality of schooling, teacher expectancy and motivation, language deprivation, culture-biased tests, nutrition, and reproductive casualty—and finds them inadequate to account for the known facts of black performance on tests of mental ability.

6. Therefore, since there is overwhelming evidence that genetic factors are the most important determinant of IQ differences within racial groups (the correlation between genotypes and phenotypes being of the order of 0.80 to 0.90), it seems most reasonable to hypothesize that "factor X" might be a racial difference in the genetic factors involved in the development of intelligence as measured by standard tests. This hypothesis seems highly likely. In terms of what is already known about human evolution, about a host of other kinds of genetic racial differences,

about the relative contributions of genetic and environmental factors to differences in mental abilities, and about the regularities in developmental changes from infancy to maturity evinced by a wide variety of psychological indices, it appears highly probable (though of course it is not rigorously proved in the sense of a logically inescapable conclusion) that genetic factors are involved to a substantial degree in the lower average IQ of American blacks. So far, I have not seen a serious attempt to adduce evidence or comprehensive argumentation based thereon to the effect that the genetic hypothesis as I have stated it is either improbable or scientifically unwarranted.

## NOTES

1. I base this estimate on the study by Burks (1928), which correlated adopted children's IQs with several indices of the quality of the home environment, including mental measurements of the adoptive parents. Since the multiple correlation of IQ and the environmental variables was .350, the standard deviation of environmental effects (assuming the *SD* of IQ is 15) would be .35 × 15 = 5.25 IQ points. But as Munsinger (1975, p. 637) has pointed out, the multiple *R* of .350 is almost certainly an overestimate of the true correlation, since multiple correlations capitalize on chance. A proper statistical correction (viz. "correction for shrinkage") for this artifact reduces the multiple *R* between IQ and the environmental variables to .255, so that the *SD* of the environmental effect would amount to .255 × 15 = 3.83 IQ points. It should be noted that this value represents an estimate not of the total environmental effect on IQ, but only the effect due to environmental variation between families, which does not include the microenvironmental variations that contribute to the nongenetic differences among children within the same family.

2. As one example, we can look at a study by sociologist Jane Mercer, which includes an exceptionally detailed rating of environments of large samples of white and black school children.

The measured environment variables were:

1. Mother's participation in formal organizations (i.e., organizations of "Anglo" society), as an index of exposure to cultural materials, values, etc. of Anglo society. Consists of five different measures.
2. Minority neighborhood vs. Anglo neighborhood.
3. Cultural barriers—e.g., fluency of mother's English, knowledge of school, etc.
4. Socioeconomic status—an index based on the occupational status of the head of household and the total years of formal education.

5. Urbanization—the extent to which child's parents were exposed to American urban society during their childhood.
6. Home ownership.
7. Individualistic achievement values—based on a composite of several questionnaires of values intended to assess the extent to which "Anglo" values had been internalized by the mother.
8. Family structure—e.g., whether child lives with both biological parents, etc.
9. Anxiety (Sarason's School Anxiety Scale)—fifteen items assessed in an interview with child.

The multiple correlation (corrected for attenuation) between these nine factors and Full Scale WISC IQ was 0.44 for blacks and 0.37 for whites, accounting for 19.6 and 13.6 percent of the IQ variance, respectively.

It is likely that these multiple correlations include more than just environmental variance. Some degree of genetic correlation is almost certainly involved in the multiple $R$ between these environmental ratings and IQ if one acknowledges the compelling evidence for a genetic component in social class IQ differences *within* racial groups. So the multiple correlations of 0.44 and 0.37 in all probability considerably overestimate the true correlation—i.e., independent of genotypes—between this set of environmental measures and the Wechsler IQ. Also, remember that a multiple correlation to some extent capitalizes on chance, and when cross-validated in another sample the same regression coefficients will yield a somewhat lower $R$. It is interesting that Mercer's $R^2$ (the proportion of variance attributable to ratings of the environment), even though it very probably contains genetic variance, does not exceed the proportion of environmental variance generally found in heritability studies of IQ based on twin and other kinship correlations, i.e., $1 - h^2$, which is about 0.2 to 0.3.

Thus these environmental indices, which all together account for about 16 percent of the IQ variance (i.e., $R^2 = 0.4^2 = .16$) *within* racial groups, if applied to the *between*-groups variance (which in Mercer's samples is $[1.3\sigma/2]^2 = .42\sigma^2$) yields $0.16 \times 0.42\sigma^2 = .067\sigma^2$. On the IQ scale (with $\sigma = 15$), this is equivalent to about four IQ points out of the average fifteen IQ points difference between the racial groups. Mercer tries to explain the total IQ difference by partialling out (statistically controlling) variables that are very likely highly correlated with parental genotypes for IQ (e.g., occupational status and education) and with race (e.g., living in a racially segregated neighborhood). So we arrive again at the conclusion that environmental indices, accounting for much if not all of the available environmental variance in IQ *within* racial groups, accounts for only about a third of the mean difference *between* the racial groups. To account for all of the racial difference in terms of environment, one must assume that the environmental variance *between* groups is much greater (by a factor of about 4)* than *within* groups, or hypothesize the influence of some nongenetic factor or factors other than those which account for the environmental component of IQ variance *within* racial groups. So far, guesses as to the nature of these environmental factors needed to fill the gap are entirely *ad hoc*. It seems to me less *ad hoc* to hypothesize that the missing factors are genetic, since genetic factors are already established as important determinants in the development of intelligence.

## REFERENCES

Burks, B.S., "The Relative Influence of Nature and Nurture upon Mental Development: A Comparative Study of Foster Parent-Foster Child Resemblance and True Parent-True Child Resemblance." *Twenty-Seventh Yearbook of the National Society for the Study of Education*, 1928, 27 (Part I), 219-316.

Mercer, J. and Brown, W.C., "Racial Differences in IQ: Fact or Artifact?" In C. Senna (Ed.) *The Fallacy of IQ*. New York: The Third Press, 1973. Pp. 56-113.

Munsinger, H., "The Adopted Child's IQ: A Critical Review." *Psychological Bulletin*, 1975, *82*, 623-659.

*Actually, the between-group variances on Mercer's environmental indices are generally less than half the within-group variances.

# Appendix III: Notes on IQ Genotype and Social Class

Christopher Jencks

## A. DISCLAIMERS

The topic for discussion is genetic differences between socioeconomic groups. I will focus exclusively on differences between such groups with respect to genotypic IQ. The arguments made with respect to IQ apply to greater or lesser extent to other partially heritable traits, such as schizophrenia, that might reasonably be expected to influence socioeconomic success, but there is not enough quantitative evidence to assess the likely magnitude of the relationships.

## B. BASIC DATA FOR ADULTS

American data on the relationship between IQ and various measures of cultural and economic status are reviewed in Appendix B of my book, *Inequality*. The magnitude of the relationship obviously depends on the kind of status one studies.

The correlation between adult IQ scores (primarily but not exclusively on military aptitude tests) and years of school completed is usually between 0.60 and 0.70 in the U.S. Correlations between adult IQ scores and occupational status depend on the system used to rank occupations. The correlation is obviously highest when each occupation's status is defined in terms of the mean IQ of its members. When this is done the correlation is usually between 0.50

and 0.60 in the U.S. It is a bit higher in Great Britain. When occupations are ranked in terms of mean earnings, the correlation with individual IQ falls to between 0.40 and 0.50 in the U.S. No English data are available yet. The best all-round index of occupational status—i.e., the index that best predicts a son's life chances—seems to be the Duncan scale, which is based on the educational requirements and economic rewards of the occupation. The correlation between an individual's IQ score in adulthood and his Duncan score is on the order of 0.50. Correlations between adult IQ and income for men between 25 and 34 average between 0.30 and 0.35. For older men, after correcting for attenuation, the figure seems to be about 0.40.

### C. CAUSAL AMBIGUITIES FOR ADULTS

There is some causal ambiguity in the relationship between adult IQ and other adult traits. Two of the three available studies suggest that the amount of schooling an individual gets will exert an appreciable influence on his or her adult IQ score. This may also be true of the type of work an individual does. This raises several related questions.

1. Are adult environments more or less uniform than juvenile environments? If they are less uniform, does this mean that adult IQ scores have lower heritabilities than juvenile scores?

2. If we assume that the same genetic factors influence both adult and juvenile test scores, and if we assume comparable heritabilities, the heritability cannot exceed the correlation between adult and juvenile scores. Indeed, the combined effects of genetic variance, between-family environmental variance, and covariance between genotype and family background cannot exceed the correlation between adult and juvenile IQ scores. This correlation is not well established for representative samples, but the observed value rarely exceeds 0.80. This seems to imply that broad heritability is 0.60 or less, since the combined effects of between-family environmental variance and covariance are likely to be at least 0.20.

   Alternatively, one could argue that the heritability of adult test scores is probably lower than the heritability of juvenile test scores, due to more variable environments. In the absence of direct evidence on the heritability of adult scores, I am inclined to accept this view.

## D. BASIC DATA FOR CHILDHOOD IQ AND ADULT SES

Given these difficulties, the only "clean" method of estimating the impact of genotype on adult success is probably to obtain correlations between juvenile scores and adult success. For educational attainment, the correlations are usually in the 0.50-0.60 range. For occupational status, they are in the 0.35 to 0.45 range. No data are available for income.

## E. CLASS DIFFERENCES IN IQ GENOTYPE: HERRNSTEIN'S SYLLOGISM

Now consider Herrnstein's syllogism, a slightly modified form of which goes as follows:

1. If genotype influences IQ, and
2. If IQ influences an individual's eventual SES, then
3. IQ genotype will influence eventual SES.
4. And since IQ genotype is inherited,
5. Children with high SES parents will tend to have more favorable IQ genotypes than children with low SES parents.

This syllogism is presented visually in the path model in Figure III-1.

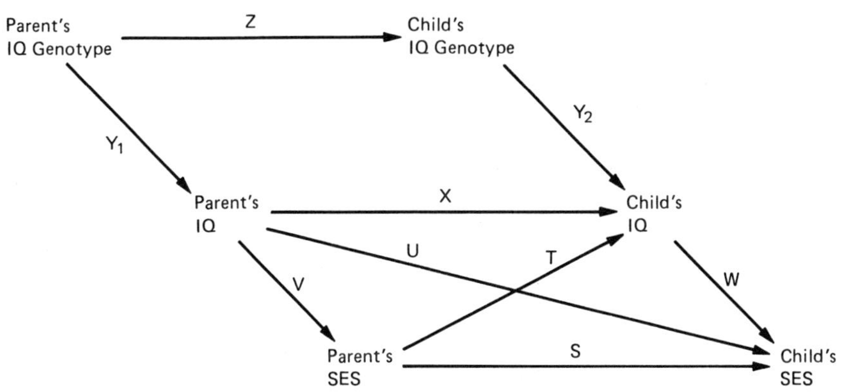


**Figure III-1.**

The convention for reading the model is that variables on the left are temporally prior to those on the right. This means that a variable can only influence variables to its right, not to its left. In addition, the model assumes that certain temporally conceivable influences are in fact negligible. Specifically, it assumes that the child's genotype depends only on the parent's genotype. Thus the parent's IQ and SES have no *direct* effect on the child's genotype, though they will be correlated with it.

Likewise, the model assumes that the parent's genotype influences the child's IQ only indirectly (via the child's IQ genotype and the parent's IQ), not directly. Finally, the model assumes that an individual's own IQ genotype influences his SES only insofar as it influences his IQ. The remaining causal paths are included in the model and are indicated by arrows that point from the independent to the dependent variable.

In terms of this model, Herrnstein's syllogism asserts that:

1. The path from IQ genotype to IQ ($Y_1$ and $Y_2$) is greater than zero.
2. The paths from IQ to SES (V and W) are greater than zero, even when other relevant factors are controlled (as they are in this model for W but not for V).
3. In light of (1) and (2) the compound path from IQ genotype to SES ($Y_1$ V or $Y_2$ W) must also be greater than zero.
4. The path from parent's IQ genotype to child's IQ genotype (Z) is greater than zero.
5. In light of (3) and (4) the correlation between parent's SES and child's IQ genotype ($ZY_1$ V) must be greater than zero.

There is no serious doubt about the heritability of genotypes (Proposition 4). Nor, despite Kamin, Putnam, and others, can I see any plausible evidence that IQ phenotype is completely independent of genotype (Proposition 1). If we are to reject Herrnstein's syllogism, then, we have to argue that an individual's IQ phenotype has no influence on his subsequent SES (Proposition 2). We know the observed correlation is positive. The only way in which the direct effect (W) could be zero, then, would be if the observed correlation were attributable to background characteristics, such as parental SES, parental IQ, or other unmeasured traits.

The most stringent way to control these background characteristics is probably to look at the effect of IQ differences within the same family. We can then measure the "true" effect of IQ on eventual SES by regressing the eventual SES difference between two children reared in the same home on the initial IQ difference between these same two children. The resulting coefficient is considerably lower than the coefficient in the general population, but it is never anything like zero.

Within-family analysis is also useful for determining whether IQ genotype can affect SES directly, or whether its effects are entirely explained by the effects of IQ phenotype on SES, as the model assumes. This can be established by comparing the regressions of eventual SES differences on IQ differences for monozygotic and dizygotic twins. The two regressions do not differ significantly. Indeed, the observed coefficient for DZ twins is actually smaller than for MZ twins. This is the opposite of what we would expect, since the differences between MZ twins presumably involve relatively more random error. This suggests that even if we could measure IQ genotype directly, we would not be able to predict adult success any more accurately than we now predict it merely from knowing phenotypic IQ.

## F. QUANTITATIVE ESTIMATES OF CLASS DIFFERENCES

Assuming that we accept Herrnstein's syllogism as theoretically correct, the interesting and important questions are: (a) whether the effects of IQ genotype on eventual socioeconomic status are large, moderate, or trivial; and (b) whether the *inheritance* of IQ genotype plays a large, moderate, or trivial role in the *inheritance* of socioeconomic status. Broadly speaking, there are two ways of trying to answer both questions.

The first approach is to use studies of adopted children. The results of these studies are summarized in *Inequality* (pp. 77-81). The second approach is to construct an internally consistent path model for two generations. This method is used in Appendix B of *Inequality*. A simplified version of these calculations, designed to yield a maximum estimate of the effect of IQ genotype on occupational status, is presented in the Appendix attached to these notes. The computations depend on the assumption that the correlation between IQ and SES is stable over time. All the evidence with which I am familiar supports this assumption.

Table III-1 summarizes what seem to me the most reasonable conclusions that can be drawn from the data currently available. Column 1 shows the correlation between IQ at age 11-14 and eventual educational attainment, occupational status (Duncan scale) and income. Rows 1-3 of Column 2 show the correlation between IQ

**Table III-1. Estimated Relationship Between Parental Characteristics, Children's Characteristics, and Children's IQ Genotype: U.S. Males**

| *Respondent's Characteristic* | *Respondent's IQ at age 11-14*[a] | *Respondent's IQ Genotype*[b] | *Same Trait in Father*[a] | *Percent Reduction in Column 3 When IQ Genotype Controlled*[c] | *Path with IQ Genotype Controlled* |
|---|---|---|---|---|---|
| Educational attainment | .50-.60 | .40-.55 | .40-.50 | 10-20 | .32-.45 |
| Occupational status | .35-.50 | .28-.45 | .40-.50 | 5-15 | .34-.48 |
| Annual income | [.25-.35] [d] | .20-.32 | [.30-.45] | 5-15 | .30-.43 |
| *Father's Characteristic* | | | | | |
| Educational attainment | .30-.40 | .20-.30 | | | |
| Occupational status | .30-.40 | .15-.25 | | | |
| Annual income | .25-.35 | .10-.20 | | | |

[a]Taken from data in *Inequality*, adjusted for various possible levels of measurement error and restriction in sample range.

[b]First value in rows 1-3 derived by multiplying first value in column 1 by 0.80 (based on Jencks' estimate of the correlation between genotype IQ and phenotypic IQ, i.e., 0.80). Second value in rows 1-3 derived by multiplying second value in column 1 by 0.90 (based on Jensen's implied correlation between genotype and phenotype). Values in rows 4-6 derived by multiplying values in rows 1-3 by 0.55, i.e., the estimated correlation between fathers' and sons' genotypes with assortative mating and some dominance.

[c]See Appendix III-A for method of derivation.

[d]No direct data available; estimated from path models.

genotype and these same outcomes. Rows 4-6 of Column 2 show the correlation of IQ genotype with these same characteristics of one's parents. Column 3 shows the correlation between fathers and sons on educational attainment, occupational status, and income. Column 4 shows the fraction of the observed correlation in Column 3 that might plausibly be attributed to inheritance of IQ genotype. The estimates in Column 4 are derived using the method given in Appendix III-A. Each column gives maximum and minimum likely values.

The one item of information not given in Table III-1 is the fraction of the correlation between parental SES and a child's IQ that is genetic in origin. Based on the available data, the figure could be anywhere from 25 to 75 percent—a rather wide margin of error.

## G. EQUALITY OF OPPORTUNITY

According to classical liberal theory, "equality of opportunity" means that an individual's eventual position in the social hierarchy should not depend on his parents' position, except insofar as this is a proxy for "native ability." Table III-1 gives some indication of the extent to which American society meets this standard. Column 3 gives us a gross measure of "inequality of opportunity," i.e., the extent to which parental position determines a child's position. Column 5 gives us an estimate of "net inequality of opportunity," i.e., the effect of parental position once the child's IQ genotype is controlled. Column 2 gives us a measure of the extent to which America is what we might call an "IQ Meritocracy." The evidence suggests that parental position and IQ genotype have roughly comparable influence. Neither explains most of the variation in educational attainment, occupational status, or annual income.

## H. COMPENSATORY OPPORTUNITY

Many people feel that belief in genetic differences between high and low SES children weakens the case for compensatory education. I cannot see why this should be true.

1. If the case for compensatory education is moral, it seems as reasonable to compensate a child for inheriting the "wrong" genes as for having parents with the "wrong" culture.
2. If the case for compensatory education is practical, it seems as likely that intervention programs will be able to raise the test scores (or life chances) of the genetically deprived as well as of the environmentally deprived.

If the determinants of cognitive skill are additive, as we often assume, the effect of a remedial program would be the same regardless of whether the student's initial problems were primarily genetic or environmental in origin. There may, of course, be interactions between the determinants of cognitive skill. In that case, remedial programs may benefit the environmentally disadvantaged more than the genetically disadvantaged. Or they may work the other way round. So far as I know, nobody has investigated this question. In the absence of evidence, all we can do is speculate.

The "standard" speculation has been that low SES children do worse on standardized tests because their parents provide them with relatively few opportunities to acquire middle class skills and information. It seems to follow that if schools give both groups comparable opportunities, the gap will disappear. This theory is clearly reasonable in some instances. If a student has been raised speaking French, for example, he will do poorly on the Metropolitan Readiness Test when he enters an English-speaking school. After a couple of years of school, however, he will usually have learned a good deal of English, and the performance gap between him and other children will have diminished sharply.

The experience of the 1960s suggests that the damage done by a poor home is usually harder to remedy than the damage done by learning the "wrong" language prior to first grade. The test score gap is temporarily reduced by preschool programs that put low SES children in school younger than high SES children, but the gap widens again once the high SES children enter school too. The test score gap persists throughout elementary school. Indeed, it may even widen slightly. It remains stable through secondary school, too.

The gap is not usually reduced by compensatory programs of the type funded under Title 1 of the Elementary and Secondary Education Act. Programs like "Sesame Street," which are available to almost everyone but are used disproportionately by middle class children, appear to widen the gap. Nor does compensatory college admission seem to affect the gap much. Low SES students presumably tend to have more favorable genotypes than high SES students with comparable SAT scores. Yet low SES students do not earn higher college grades than high SES students with comparable SAT scores. In general, the partial correlation of grades with SES tends to be zero once initial test scores are controlled.

The critical question about test score differences between high and low SES students is not, then, whether they are genetic or environmental in origin, but whether they can be eliminated by some specific intervention. If differences are ineradicable, it makes little difference whether they are genetic or environmental in origin. The same is true if they are eradicable. The mere fact of their being genetic or environmental does not tell us whether they are eradicable or not.

## I. THE EFFECTS OF USING TEST SCORES TO PREDICT PERFORMANCE

The correlation between class background and IQ is generally higher than the correlation between class background and school or college grades. Thus if a college uses IQ (or SAT) scores to select among applicants, it will end up excluding more low SES students than if it admits everyone and then excludes those whose grades fall below a comparable cutoff point. Suppose, for example, that a college has 2,000 applicants, half "high" SES and half "low." Suppose it has room for a total enrollment of 5,000 students in all four years. Suppose that with open admissions SES correlates 0.30 with SAT scores and 0.20 with freshman grades—fairly typical figures. Suppose, too, that under open admissions SAT scores correlate 0.60 with freshman grades, and that the average freshman has a grade point average of 2.0, an S.D. of 1.0, and a binomial distribution (i.e., 1/16 Fs, 1/4 Ds, 3/8 Cs, 1/4 Bs, 1/16 As).

If the college uses open admissions and flunks out the bottom half of the class at the end of the freshman year, it will end up with 3,000 upperclassmen, 60 percent of whom will be high SES; 40 percent will be low SES. The mean GPA (assuming the same standards are applied to upperclassmen as freshmen) will be 2.75. If the college admits the applicants with the highest SAT scores and then flunks nobody, it will admit 1,250 applicants: 62 percent will be high SES, while 38 percent will be low SES. The predicted grade point average will be 2.45. The use of SATs rather than freshman grades to select upperclassmen will thus have three effects:

1. It will reduce the percentage of low SES upperclassmen from 40 to 38.
2. It will reduce the grade point average for upperclassmen from 2.75 to 2.45 by eliminating some who could do the work and accepting some who will do poorly.
3. It will reduce the cost per graduate, since it costs less than $100 apiece to test and reject applicants, whereas it costs several thousand dollars apiece to admit students and flunk them out a year later.

A fourth probable effect is that the use of SAT scores will make the freshman year pleasanter for those admitted. Students will usually be presumed competent, simply because they are admitted. Open admissions plus a high failure rate makes the freshman year a survival test.

The foregoing logic also applies to selecting applicants for high status jobs.

## APPENDIX III-A

**Path Models of Maximum Likely Effects of IQ Genotype in Transmission of Status**

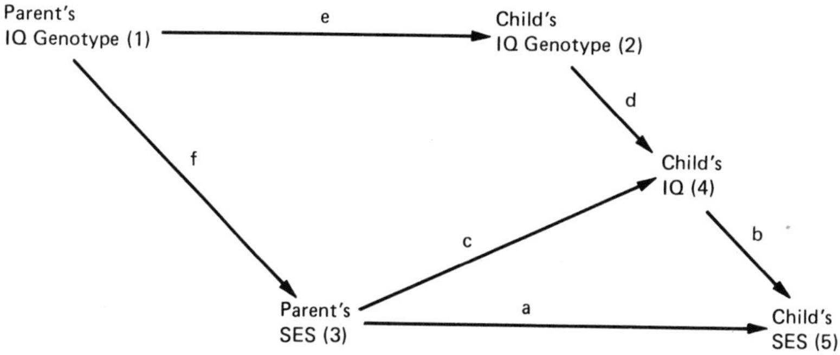


**Figure III-A-1.**

1. *Observed Correlation* of parent's and child's status

   $r_{5,3} = a + br_{4,3} = a + bc + bdef$

2. *Effect* of parental status on child's status, controlling parent's IQ genotype

   $b_{5,3.1} = a + bc$

3. *Percentage of correlation accounted for by genotype* =

   $$\frac{bdef}{a + bc + bdef}$$

4. *Constraint:* Correlation of genotype and status must be equal for parents and children, i.e., $r_{3,1} = r_{5,2}$

To maximize $\frac{bdef}{a + bc + bdef}$ we want maximum values for b, d, e, and f.

If we define status in terms of occupational status, the maximum reported value of $r_{5,4}$ is Bajema's (1968) report of 0.46. Correcting for measurement error, assume $r_{5,4} = 0.50$. For d, we use Jensen's estimate of 0.90. For e, we use 0.55.

In addition, $r_{5,3} = 0.405$ from Blau and Duncan (1967). Correcting for measurement error, my best estimate is $r_{5,3} = 0.483$. For $r_{4,3}$, we can use 0.357 (from *Inequality*, Table B-2).

Then we have:

(1) $r_{5,4} = b + ar_{4,3} = 0.50 = b + 0.357a$

(2) $r_{5,3} = a + br_{4,3} = 0.483 = a + 0.357b$

Solving, we get a = 0.349 and b = 0.375

Then we have:

(3) $r_{4,3} = c + def$

but d = 0.90 and e = 0.55, and $r_{4,3} = 0.357$, so

(3a) $0.357 = c + 0.495f$

or

(3b) $c = 0.357 - 0.495f$

in addition,

(4) $r_{3,1} = r_{5,2} = f = ar_{3,2} + br_{4,2} = afe + bd + bcfe$

Substituting the values of a and b we obtained from equations (1) and (2), and the maximum values of d and e derived from the literature, and then simplifying, we have

(4a) $0.808f - 0.206cf = 0.338$

Substituting (3b) into (4a) and simplifying, we get:

(4b) $0.102f^2 + 0.734f - 0.338 = 0$

Which yields f = 0.434.

And so,

$$\text{(3c) } c = 0.357 - (0.495)(0.434) = 0.142$$

Thus we get $r_{5,3.1} = a + bc = 0.402$

$$\text{and } \frac{bdef}{a + bc + bdef} = \frac{0.081}{0.402 + 0.081} = 0.167$$

Hence a maximum of 16.7 percent of the observed correlation between parental status and child's status is attributable to the child's advantages with respect to IQ genotype; a minimum of 83.3 percent is due to other factors. These other factors include both genetic advantages that produced "desirable" traits other than IQ (e.g., a white skin) and environmental advantages.

Note that the correlation between parental SES and a child's IQ is 0.357 in this model, while the direct path after controlling parental genotype is 0.142. This implies that (.357 — .142)/.357 = 60.2 percent of the correlation is attributable to genetic factors, and 39.8 percent to environmental factors.

Note, too, that the correlation between a child's IQ genotype and parental status is (0.55)(0.434) = 0.24. This means that there is some covariance between genetic and environmental determinants of IQ. (The minimum estimate would be (2)(.142)(.90)(.24) = 6 percent of the total IQ variance.) It also means that the correlation between IQ and genotype ($r_{4,2}$) exceeds the direct path. In this model, $r_{4,2}$ = (.90) + (.142)(.434)(.55) = .934, and $r^2_{4.2}$ = .872. This is a *very* high estimate of the heritability of IQ, and cannot, I believe, be reconciled with other data. It is presented here only to illustrate a method and to show that even on this assumption, IQ genotype does not play a central role in social stratification.

❋

# References


Baker, J.R. *Race.* New York: Oxford University Press, 1974. (See especially pp. 438-441, 468-471, 495-497.)

Bereiter, C. "The Future of Individual Differences." *Harvard Educational Review* 39 (2) (Spring 1969): 162-166.

Cavalli-Sforza, L.L. and Bodmer, W.F. *The Genetics of Human Populations.* San Francisco: W.H. Freeman, 1971. (See especially pp. 24-36.)

Cronbach, L. *Essentials of Psychological Testing.* New York: Harper and Row, 1970.

Dobzhansky, T. *Mankind Evolving.* New Haven: Yale University Press, 1962.

Dobzhansky, T. "Differences Are Not Deficits." *Psychology Today* 7 (7) (December 1973): 97-101.

Jencks, C. *Inequality.* New York: Basic Books, 1972. (See especially Appendix B, "Path Models of Intergenerational Mobility," pp. 320-350.)

Jensen, A.R. *Educability and Group Differences.* New York: Harper & Row, 1973. (See especially pp. 1-13, 128-129, 151-154.)

Jensen, A.R. "The Differences Are Real." *Psychology Today* 7 (7) (December 1973): 80-86.

Layzer, D. "Heritability Analyses of IQ Scores: Science or Numerology." *Science* 183 (29 March 1944): 1259-1266.

Lerner, I.M. *Heredity, Evolution and Society.* San Francisco: W.H. Freeman, 1968.

Loehlin, J.; Lindzey, G.; and Spuhler, J.N. *Race Differences in Intelligence.* San Francisco: W.H. Freeman, 1975. (See especially the chapter on "Heritability.")

Mather, K. *Human Diversity.* New York: Free Press, 1964. (See especially pp. 75-79.)

McClearn, G.E. "Behavioral Genetics: An Overview." *Merrill-Palmer Quarterly of Behavior and Development* 14 (1) (1968): 9-24.

Provine, W.B. "Geneticists and the Biology of Race Crossing." *Science* 182 (23 November 1973): 790-796.

Rice, B. "The High Cost of Thinking the Unthinkable." *Psychology Today* 7 (7) (December 1973): 89-93.

Thoday, J.M. "Review of *Educability and Group Differences.*" *Nature* 245 (26 October 1973): 418-420.

# Index